THE SCHOOL MATHEMATICS PROJECT

When the S.M.P. was founded in 1961, its objective was to devise radically new mathematics courses, with accompanying G.C.E. syllabuses and examinations, which would reflect, more adequately than did the traditional syllabuses, the up-to-date nature and usages of mathematics.

The first stage of this objective is now more or less complete. *Books 1–5* form the main series of pupils' texts, starting at the age of 11+ and leading to the O-level examination in 'S.M.P. Mathematics', while *Books 3T, 4* and *5* give a three-year course to the same O-level examination. (*Books T* and *T4*, together with their Supplement, represent the first attempt at this three-year course, but they may be regarded as obsolete.) *Advanced Mathematics Books 1–4* cover the syllabus for the A-level examination in 'S.M.P. Mathematics' and in preparation are five (or more) shorter texts covering the material of various sections of the A-level examination in 'S.M.P. Further Mathematics'. There are two books for 'S.M.P. Additional Mathematics' at O-level. Every book is accompanied by a Teacher's Guide.

For the convenience of schools, the S.M.P. has an arrangement whereby its examinations are made available by every G.C.E. Examining Board, and it is most grateful to the Secretaries of the eight Boards for their cooperation in this. At the same time, most Boards now offer their own syllabuses in 'modern mathematics' for which the S.M.P. texts are suitable.

By 1967, it had become clear from experience in comprehensive schools that the mathematical content of the S.M.P. texts was suitable for a much wider range of pupil than had been originally anticipated, but that the presentation needed adaptation. Thus it was decided to produce a new series, *Books A–H*, which could serve as a secondary school course starting at the age of 11+. These books are specially suitable for pupils aiming at a C.S.E. examination; however, the framework of the C.S.E. examinations is such that it is inappropriate for the S.M.P. to offer its own examination as it does for the G.C.E.

The completion of all these books does not mean that the S.M.P. has no more to offer to the cause of curriculum research. The team of S.M.P. writers, now numbering some thirty school and university mathematicians, is continually testing and revising old work and preparing for new. At the same time, the effectiveness of the S.M.P.'s work depends, as it always has done, on obtaining reactions from active teachers—and also from pupils—in the classroom. Readers of the texts can therefore send their comments to the S.M.P. in the knowledge that they will be warmly welcomed.

Finally, the year-by-year activity of the S.M.P. is recorded in the annual Director's Reports which readers are encouraged to obtain on request to the S.M.P. Office at Westfield College, University of London, Kidderpore Avenue, London NW3 7ST.

ACKNOWLEDGEMENTS

The principal authors, on whose contributions the S.M.P. texts are largely based, are named in the annual Reports. Many other authors have also provided original material, and still more have been directly involved in the revision of draft versions of chapters and books. The Project gratefully acknowledges the contributions which they and their schools have made.

This book—*Teacher's Guide for Book B*—has been written by

        J. K. Brunton   W. Mrozowski
        Joyce Harris   Margaret Wilkinson
        K. Lewis

and edited by P. G. Bowie assisted by Elizabeth Evans.

Many other schoolteachers have been directly involved in the further development and revision of the material and the Project gratefully acknowledges the contribution which they and their schools have made.

We would especially thank Dr J. V. Armitage for the advice he has given on the fundamental mathematics of the course.

The drawings at the chapter openings in *Book B* are by Penny Wager.

The Project is grateful to the Royal Mint for supplying the photographs of the coins which appear in Chapter 3.

We are much indebted to the Cambridge University Press for their cooperation and help at all times in the preparation of this book.

THE SCHOOL MATHEMATICS PROJECT

TEACHER'S GUIDE FOR BOOK B

CAMBRIDGE
AT THE UNIVERSITY PRESS  1969

Published by the Syndics of the Cambridge University Press
Bentley House, 200 Euston Road, London NW1 2DB
American Branch: 32 East 57th Street, New York, N.Y.10022

© Cambridge University Press 1969

Library of Congress Catalogue Card Number: 68-21399

ISBN: 0 521 07223 9

First published 1969
Reprinted 1971 1972

Printed in Great Britain
at the University Printing House, Cambridge
(Brooke Crutchley, University Printer)

Preface—Book B

This is the second of eight books designed to cover a course suitable for those who wish to take a CSE Examination on one of the reformed mathematics syllabuses.

The material is based upon the first four books of the O-level series, SMP *Books 1—4*. The connection is maintained to the extent that it will be possible to change from one series to the other at the end of the first year or even at a later stage. For example, having started with *Books A* and *B*, a pupil will be able to move to *Book 2*. Within each year's work, the material has been entirely broken down and rewritten.

The differences between this Main School series and the O-level series have been explained at length in the Preface to *Book A* as have the differences between the content of these two SMP courses and that of the more traditional text.

In this book, *Book B*, and in the remainder of the series, metric units are used. Money is in pounds and new pence; length in metres and also centimetres and kilometres; weight is in grammes and kilogrammes. Measures of time and angle remain unchanged.

The Prelude introduces work that is preliminary to the tessellation and area chapters. It is concerned with patterns that fill the plane. The chapter on tessellation then concentrates upon the angle and side properties necessary in the tessellating polygons while the area chapter concentrates upon measurement by comparison with any unit area.

There are two other geometry chapters. One concerns the application of the idea of angle to bearing and direction. The other introduces the study of transformations. In later books we shall meet transformations which preserve distance and shape, as we did in a preliminary way in the symmetry chapter of *Book A*. We shall also meet transformations such as enlargement and shearing, but in this book we consider a transformation that preserves neither shape nor size, but only properties of containment, order and incidence.

The decimals chapter is really about measurement, the difficulty of obtaining and expressing accurate measurements and of stating the degree of accuracy. It relies heavily upon the experience of place value developed in the number base chapter of *Book A*. The number base chapter of this book concentrates less upon place value than upon the applications of binary arithmetic.

The fractions chapter continues the efforts started in *Book A*, to make fractions more completely understood without stressing manipulation. A chapter on directed numbers introduces the signs (+) and (−) as generaliza-

Preface—Book B

tions of 'above, below', before, 'after', etc. Directed numbers as shifts will be discussed in *Book C*.

There are two algebra chapters: one on the meaning of letters in mathematical context; (equations are not discussed; the emphasis is upon the meaning of a symbol 'for all x in the set S...' and the formation of simple formulae); the other on relations, mappings and ordered pairs.

Preface—Teacher's Guide for Book B

This book contains the pages from pupil's *Book B* interleaved with comments and answers to questions. The comments have three aims in mind. They try to show the purpose of new and unusual topics and fresh approaches to old topics; they provide occasional notes and references on the background to the mathematics and they discuss how the contents of *Book B* can best be presented to classes.

In the pupil's book we tried to produce a mathematics text which pupils could use with only occasional help from their teachers. We hope that the books will initiate activities and investigations beyond the immediate problem, and sometimes outside the classroom itself, while indicating the lines of development to which the pupils can eventually return.

The background notes give references to those sources which we feel are of the greatest direct value, though there are many books of more general interest. We hope that teachers who have used this guide in conjunction with the pupil's book will send us comments and criticisms. Only in this way can we be sure that the guide will continue to serve its purpose.

Publisher's note

This book consists of pupil's *Book B* interleaved with pages of answers and commentary for the teacher; teacher's pages are distinguished by a red line down the margin.

The book is numbered in red throughout with the prefix T, e.g. T39 For ease of reference pupil's pages also retain the black number that appears on the identical page in *Book B*. The teacher's numbering is used in the list of contents and the index, and the list of contents also repeats the pupil's number for the beginning of each chapter.

Contents

		page	
Preface		*page*	v
Prelude		T1	1
Tiling patterns, *T1*			
1 Letters for numbers		T14	6
Patterns among numbers, *T14*; sets and subsets, *T18*; formulae, *T22*			
2 Tessellations		T28	13
Pattern, *T28*; symmetry, *T42*			
3 Decimals		T53	23
Measurement, *T54*; standard units of length, *T60*; addition and subtraction shown on the abacus, *T64*; decimal coinage, *T68*; rounding off and significance, *T72*			
Interlude		T79	35
4 Area		T82	36
Comparison of area, *T82*; measurement of area, *T92*; areas of rectangles, *T102*			
5 Comparison of fractions		T115	51
Representing fractions, *T116*; equivalent fractions, *T120*; comparing fractions, *T128*; Farey sequences, *T138*			
Revision exercises		T142	63
6 Angle		T152	68
Fixing a position, *T152*; the clock-ray method, *T154*; bearings, *T156*; radar, *T174*			
7 Relations		T180	81
Family relationships, *T180*; mappings, *T190*; ordered pairs, *T196*			
8 Binary and duodecimal bases		T200	90
Revision, *T200*; base twelve, *T204*; base two, *T212*; base two fractions, *T226*			
9 Statistics		T234	106
A survey, *T234*; bar charts, *T236*; pie charts, *T240*; pictograms, *T242*; projects, *T244*			
Interlude		T252	116

Contents

10 Directed numbers T254   *page* **117**
  Reference points, T254; directed numbers, T260; the number line, T264

11 Topology T270   **125**
  Topological transformations, T270; nodes, T282; arcs and regions, T286; traversable networks, T288; inside or outside, T294; colouring regions, T298

Puzzle corner T314   **144**

Revision exercises T322   **147**

Bibliography T336

Index T337

Equipment and materials

The following items are used so frequently that they can be considered as basic equipment for the course:

Sets of rulers, compasses, protractors, scissors. Thin card, tracing paper, plain paper, centimetre squared paper, isometric paper.

Some chapters require special equipment. These are:

Prelude Set of nine-by-nine pinboards, rubber bands for pinboards. Spotty paper.

Chapter 2 Coloured gummed paper.

Chapter 3 Set of spike abaci and E.S.A. adjustable spike abacus for demonstration.

Chapter 4 Glue. String.

Chapter 6 Clockface, compass points on floor in maths room.

Chapter 8 Set of spike abaci and E.S.A. adjustable spike abacus for demonstration. Device to Illustrate the Binary System: 5 flashlamp bulbs and holders, 5 switches, wire, battery. Examples of computer tape. Packets of punched cards.

Chapter 9 Set of dice.

Chapter 10 Wall height measurer in maths room.

Much of the work can be better illustrated with the aid of a magnetic board and an overhead projector. A set of transparencies for overhead projectors, to accompany *Book B*, is published by Cambridge University Press (ISBN 0 521 08006 1).

Prelude

TILING PATTERNS

The Prelude is designed to stimulate interest in geometrical patterns and to provide some informal experiences which will form a background for work in the chapters on tessellation and area.

Experiment 1

Pupils should be allowed to decide for themselves how to obtain as many triangles as possible. (Some may cut out at random!) When they have completed Experiments 1 and 2, they should be able to appreciate that the maximum number of triangles can be obtained by starting with a tiling pattern.

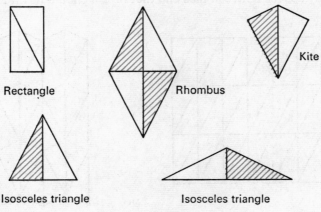

Fig. A

Some geometrical shapes are shown in Figure A; the shading indicates that the triangle has been turned over. These, however, may not be the shapes which are considered to be interesting. Some pupils may place

Prelude

triangles so that they are not 'edge to edge'; others may place one triangle on top of another. Three examples are shown in Figure B. There is an opportunity here to revise work on symmetry.

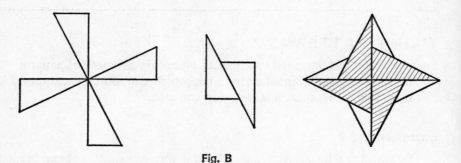

Fig. B

Two possible tiling patterns are shown in Figure C. At this stage, there is no need to insist on the tiles being arranged in an organized pattern; this will be discussed in Chapter 2. It is important that there should be no empty spaces between tiles and that no tile should lie on top of another.

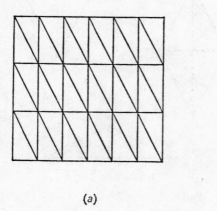
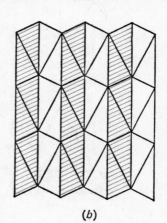

(a) (b)

Fig. C

Prelude

TILING PATTERNS

A kitchen or bathroom floor is sometimes covered with tiles all of the same shape and size. The tiles do not overlap and are fitted together without gaps where dirt could collect.

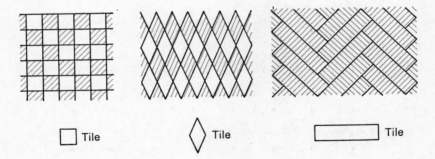

Experiment 1

Right-angled triangle

Equipment: tracing paper, thin card, scissors.

Trace this shape carefully. Use your tracing to help you to cut out triangles having the same shape and size. Try to get as many as possible from your card.

Make some interesting shapes with your triangles. Can you use your triangles to make a tiling pattern?

Prelude

Experiment 2

Equipment: tracing paper, thin card, scissors.

Repeat Experiment 1 for each of the following shapes.

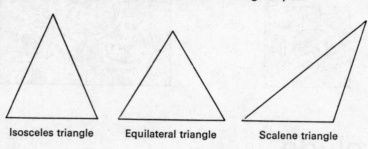

Isosceles triangle   Equilateral triangle   Scalene triangle

Draw a triangle of any shape you choose. Can you use it to make a tiling pattern? Do you think it is possible to form a tiling pattern from any triangle?

Experiment 3

Equipment: a nine-by-nine pinboard, spotty paper.

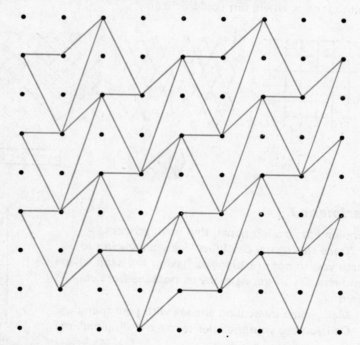

Pinboards can be used to make tiling patterns.

Tiling patterns

Experiment 2

Some geometrical shapes are shown in Figure D. No new shapes can be obtained by turning over the isosceles or equilateral triangle; they have line symmetry.

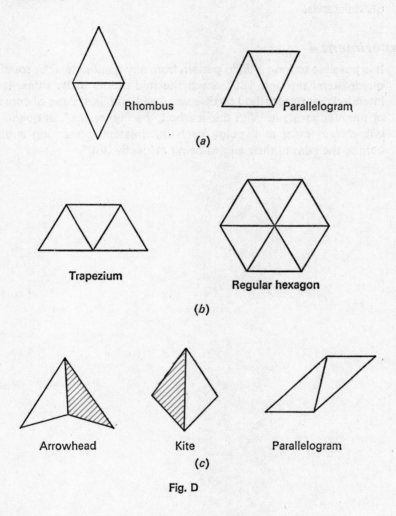

Rhombus

Parallelogram

(a)

Trapezium

Regular hexagon

(b)

Arrowhead

Kite

Parallelogram

(c)

Fig. D

A tiling pattern can be formed from any triangle.

Prelude

Experiment 3

This experiment offers a further opportunity for pupils to become familiar with the names for special quadrilaterals. There are more possible different tessellations for rectangles, squares, parallelograms, etc., than for irregular quadrilaterals.

Experiment 4

It is possible to form a tiling pattern from any quadrilateral by rotating the quadrilateral through 180° about the mid-points of its sides. It is not intended that this method be discussed with pupils, unless, of course, one of them suggests it. With this method, the corners of four quadrilaterals will always meet at a point, each quadrilateral presenting a different corner, the sum of their angles being evidently 360°.

Tiling patterns

Form tiling patterns from each of the following quadrilaterals. (They are all quadrilaterals, but some of them have some regularity and these are given special names.) Record your results on spotty paper.

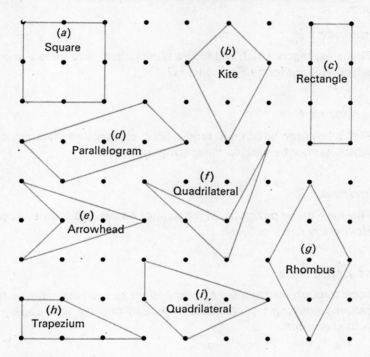

Experiment 4

Here is a tiling pattern formed from quadrilaterals.

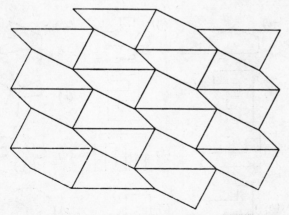

Trace it and colour all the tiles that are this way up.

Prelude

Do you think that any quadrilateral can be used to make a tiling pattern? Choose a difficult quadrilateral and find out whether a tiling pattern can be formed from it.

Experiment 5

Find a pentagon which will form a tiling pattern. Are there any pentagons which will not form tiling patterns?

Experiment 6

Find a hexagon which will form a tiling pattern. Are there any hexagons which cannot be used to make tiling patterns?

Experiment 7

There are many polygons which cannot be used to form a tiling pattern. How many can you find?

Class project

Look through newspapers and magazines to see how many actual tiling patterns you can find. Make a collection of them and display them in your classroom.

Experiment 8

Find out which of these shapes can be used to form tiling patterns.

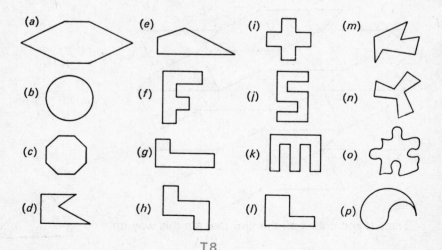

Tiling patterns

Experiment 5

This problem can be solved with the aid of a pinboard. Tiling patterns can be formed from each of the pentagons shown in Figure E. Tiling patterns cannot be formed from regular pentagons nor from, for example, those shown in Figure F.

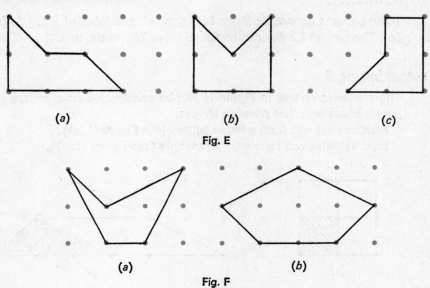

Fig. E

Fig. F

Experiment 6

Tiling patterns can be formed from regular hexagons; also from the hexagons shown in Figure G (i) but not from those shown in Figure G (ii).

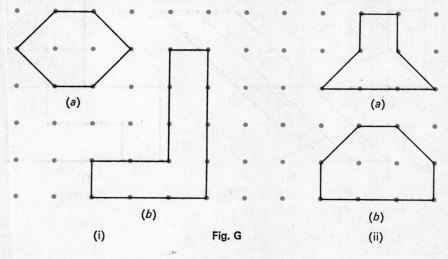

Fig. G

Prelude

Experiment 7

It is important to realize that only a minority of shapes can be used to form tiling patterns.

Experiment 8

Tiling patterns cannot be formed from the shapes labelled (*b*), (*c*), (*m*), (*p*). The pattern for (*n*) can be found on p. T40 of the text.

Experiment 9

The examples shown in Figure H can be constructed on a square pinboard. There are other possible shapes.

Four squares will form a larger square (see Figure H (*a*)).
Four triangles will form a larger triangle (see Figure H (*b*)).

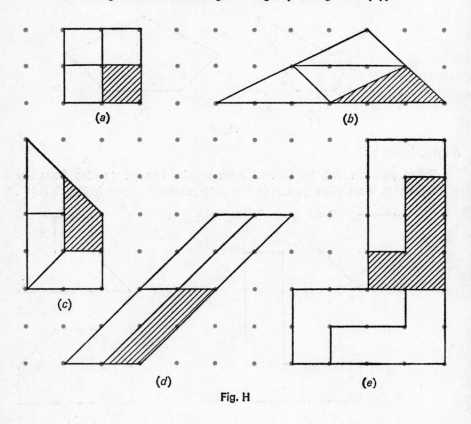

Fig. H

Tiling patterns

Four of the trapeziums shown in Figures H (c) and I (b) will form a larger trapezium.
Four parallelograms will form a larger parallelogram (see Figure H (d)).
Four rhombuses will form a larger rhombus (see Figure I (a)).
Four of the pentagons shown in Figure I (c) will form a larger pentagon.
Four of the hexagons shown in Figure H (e) will form a larger hexagon.

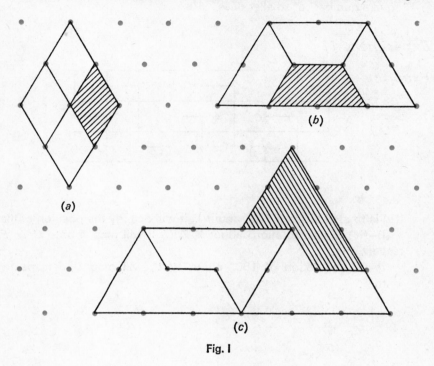

Fig. I

The experiment is included here in order to develop dexterity in fitting shapes together. It could, if desired, be used as a starting point for work on similarity and enlargement.

It would, of course, be possible to form tiling patterns from all the shapes in Figures H and I.

Prelude

Experiment 10

Larger equilateral triangles can be made from 9, 16, 25, 36, ..., of the small triangles.

For all polygons having the property mentioned in Experiment 9, it is possible (though not always easy) to form a larger similar figure from any 'square number' of smaller ones.

Experiment 11

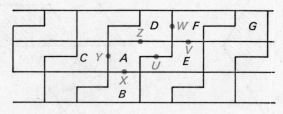

Fig. J

If A is rotated through 180° about X, it will occupy the position of tile B.

(i)–(iv) Similar rotations about Y, Z, U, V will map A onto C, D, E, G respectively.

(i)–(ii) A rotation of 180° about W, V, will map D, E respectively onto F.

Tiling patterns

Experiment 9

It is possible to make a larger equilateral triangle with four equilateral triangles of the same size and shape.
   Find other shapes which have this property. Can you form tiling patterns from all of these shapes?

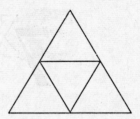

Experiment 10

Figures which have the same size and the same shape are called *congruent* figures.
   If you have a large supply of congruent equilateral triangles, you can make a larger equilateral triangle from 4 triangles. Can you make a larger equilateral triangle with any other number of triangles?
   Experiment with other shapes.

Experiment 11

Tiles A and B together form a shape with rotational symmetry.
Find the centre of rotational symmetry.
   How can tile A be moved about this centre so that it occupies the position of tile B?
   Can tile A be moved in this way so that it occupies the position of: (i) tile C; (ii) tile D; (iii) tile E, (iv) tile G? Where are the centres of symmetry?
   How can: (i) tile D; (ii) tile E be moved so that it occupies the position of tile F?
   Have you found any other tiling patterns in which two tiles could be moved to each other's position in this way?

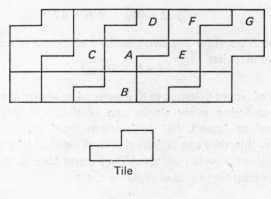

Tile

T13

1. Letters for numbers

1. PATTERNS AMONG NUMBERS

(a)

These arrangements of dots both represent the rectangle number 8. They show that
$$4 \times 2 = 2 \times 4.$$

Are the following correct

    (i) $9 \times 17 = 17 \times 9$;

    (ii) $5 \times 8 = 8 \times 5$;

    (iii) $3 \times 405 = 405 \times 3$?

Let *a* and *b* stand for any two members of the set of counting numbers. Is it always true that
$$a \times b = b \times a?$$

The use of letters helps us to say something about *all* the members of a set without listing every single one of them. In this case, they are 'substitutes' or 'stand ins' or 'understudies' for counting numbers. (Because of this, they can be joined with signs like '+' and '×' and '=' as though they really were numbers.) They stand for counting numbers until a particular counting number replaces them.

1. Letters for numbers

In *Book A*, there were occasions when letters were used in place of numbers in making general statements. For example, when referring to the sizes of angles in a triangle.

$$a° + b° + c° = 180°$$

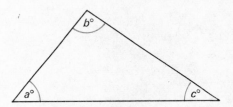

Such steps can be taken by most pupils without special explanation. Explicit reference is made to them in this short chapter, but, even here, we consider that no great emphasis need be placed upon them. Algebra is a language and it is best learnt, we feel, through continual use in situations where its usefulness is obvious. We pause here only to draw attention to what is happening, not to give an opportunity for drill.

There is one point that should perhaps be mentioned. It may be that some pupils will be familiar with algebraic statements that are true for only one or two numbers. (For example, they may have solved equations written as $4+x = 7$ or as $4+\square = 7$.) The statements of this chapter are true for all elements of the domain (in these cases, the counting numbers) and there can be no question of 'solving' anything. The difference between a statement that is true for *some* values of x and another that is true for *all* values of x need not be discussed here unless the question is raised. Examples are:

$x+6 = x^2$ true for two values of x;

$x+6 = 99$ true for one value of x;

$x+6 = \frac{1}{2}(2x+12)$ true for all values of x.

(The last equation is called an identity.)
'Equations' and methods of finding their solution sets are discussed much later in the course.

Letters for numbers

1. PATTERNS AMONG NUMBERS

The properties of numbers that are mentioned here, i.e. the commutative property ($a \times b = b \times a$) and the distributive property

$$a \times (b+c) = (a \times b) + (a \times c),$$

are only used to illustrate the way in which letters can be employed to stand for elements in the set of numbers. Their properties will be discussed in later chapters and no effort should be made to learn them now.

(b)

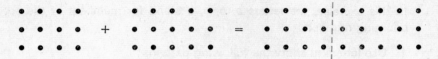

The dots show that
$$12 + 15 = 27.$$

From the arrangements of the dots,
$$(3 \times 4) + (3 \times 5) = (3 \times 9).$$

Noticing the red line
$$(3 \times 9) = 3 \times (4 + 5),$$

so
$$(3 \times 4) + (3 \times 5) = 3 \times (4 + 5).$$

Write down a similar expression for each of the following arrangements of dots:

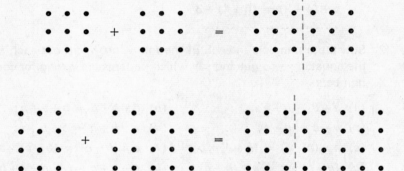

Set down an arrangement of dots for each of the following:
$$(2 \times 5) + (2 \times 7) = 2 \times (5 + 7);$$
$$(9 \times 1) + (9 \times 8) = 9 \times (1 + 8).$$

The arrangement of dots shows us one use of brackets. The numbers set out with the signs and brackets form a pattern. Is the pattern the same for all counting numbers? Can you use letters to show this pattern?

If a, b and c are members of the set of counting numbers, then this pattern is shown as
$$(a \times b) + (a \times c) = a \times (b + c).$$

T17

Letters for numbers

Exercise A

In this exercise, the letters *a*, *b*, *c*, *d* stand for members of the set of counting numbers.

1 Use letters to show the following patterns:

(a) $6+12 = 12+6$
 $1+3 = 3+1$
 ....

(b) $2 \times 3 = 3 \times 2$
 $10 \times 9 = 9 \times 10$
 ....

(c) $(2+3)+4 = 2+3+4$
 $(5+7)+8 = 5+7+8$
 ....

(d) $(5 \times 3) \times 6 = 5 \times (3 \times 6)$
 $(6 \times 1) \times 2 = 6 \times (1 \times 2)$
 ....

(e) $(3+4) \times 2 = (3 \times 2)+(4 \times 2)$
 $(7+2) \times 3 = (7 \times 3)+(2 \times 3)$
 ....

(f) $8+(6-2) = (8+6)-2$
 $5+(4-3) = (5+4)-3$
 ....

2 Substitute numbers in each of the following. Work out each side of the equations you get and say which patterns are wrong for counting numbers.

(a) $a+b = b+a$;
(b) $a+b+b = b+a+b$;
(c) $a-b = b-a$;
(d) $a-b+b = b-b+a$;
(e) $a \times (b+c) = (b+c) \times a$;
(f) $a \times (b-c) = (b-c) \times a$;
(g) $(a-b)-c = a-(b-c)$;
(h) $a+b+c+d = d+c+b+a$.

3 Make up some letter patterns like those of Question 2. Write down five of them, of which three are correct for counting numbers and two are wrong.

2. SETS AND SUBSETS

2.1 Non-numerical sets

We have been talking about sets of numbers. The word 'set' can be used for collections of other things besides numbers. For example: the set of people who live in your house; the set of people who have sailed round the world alone; the set of chairs in your classroom; the set of points that make a certain line of symmetry; the set of objects in your pocket.

Patterns among numbers

Exercise A

1. (a) $a+b = b+a$; (b) $a \times b = b \times a$;
   (c) $(a+b)+c = a+b+c$; (d) $(a \times b) \times c = a \times (b \times c)$;
   (e) $(a+b) \times c = (a \times c) + (b \times c)$; (f) $a + (b-c) = (a+b) - c$.

2. (c), (g) are not generally true for any sets of numbers, certainly not for counting numbers.

2. SETS AND SUBSETS

The chapter on relations later in this book starts with a section on non-numerical relations. It is as a foundation for that chapter that non-numerical sets are discussed here. It is often stated that sets are the basis of mathematics. This is because they are used in building up ideas of algebraic structure and only partly because they have proved useful in the sort of problems which involve Venn diagrams and Boolean algebra. We shall concentrate upon their use in the development of the idea of a relation. We shall discuss Venn diagrams later in the course.

Letters for numbers

Exercise B

2 (a) Symbols for the units of the new British Decimal Currency.
   (b) Suits of a pack of cards.   (c) The vowels.
   (d) The senses.   (e) Months with thirty days.

3 Only (c) is false.

Exercise C

1 {1, 2, 3}; {1, 2, 4}; {1, 3, 4}; {2, 3, 4}.

2 (a) Yes; provided that by 'odd numbers' is meant positive odd numbers.
   (b) No; because 'one' is not a rectangle number.
   (c) Yes.

Sets and subsets

Exercise B

1 List the members of the following sets:
   - (a) {the colours of a set of traffic lights};
   - (b) {the subjects on your timetable};
   - (c) {the days of the week};
   - (d) {the letters of your surname};
   - (e) {the five continents}.

2 Give a description which defines the following sets:
   - (a) {£, p};
   - (b) {hearts, clubs, diamonds, spades};
   - (c) {a, e, i, o, u};
   - (d) {sight, hearing, smell, touch, taste};
   - (e) {September, April, June, November}.

3 Are the following statements true or false?
   - (a) A square is a member of the set of polygons.
   - (b) The Earth is a member of the set of planets.
   - (c) An oak is a member of the set of flowers.
   - (d) Tennis is a member of the set of sports.
   - (e) Manchester is a member of the set of cities of England.

4 You are a member of your family, generation, school, the set of people who possess a pencil or pen. Name six other sets of which you are a member.

2.2 Subsets

When the members of one set A are all taken from another set B, then A is called a subset of B. Your family is a subset of the set of people who live in your road. The cars you have seen on the roads form a subset of all the cars in the world. Even numbers are a subset of the set of counting numbers. {2, 3, 5} is a subset of {1, 2, 3, 4, 5, 6}.

Exercise C

1 Make a list of all the subsets of {1, 2, 3, 4} which have three members.

2 Which of the following are true?
   - (a) {odd numbers} is a subset of {counting numbers};
   - (b) {square numbers} is a subset of {rectangle numbers};
   - (c) {multiples of ten} is a subset of {rectangle numbers}.

Letters for numbers

3 Write down the set which is the intersection of the sets

$A = \{2, 4, 6, 8, 10\}$ and $B = \{2, 4, 8, 16, 32, 64\}$.

Is A a subset of B? Is the intersection set a subset of (i) A, (ii) B?

4 There are many subsets within a family. For example {married women}, {uncles}, {people under the age of 11}. Write down five more subsets.

Some of the subsets may intersect. For example, the subset {people under the age of 11} might intersect with the subset {cousins}. Make a list of three pairs of subsets that could not possibly intersect.

Summary

The word 'set' can be used for a collection of any objects providing we can decide whether or not any particular object belongs to that collection. There must be a list of members or a clear description of the collection.

When members of one set, A, are all members of another set, B, then A is a subset of B.

3. FORMULAE

The patterns mentioned in Section 1 were very general. They applied to all counting numbers.

As you found in Book A, there are many patterns which apply only to subsets of the counting numbers.

The questions in Exercise D show you how to look for these patterns and how to describe them by means of a general expression or formula.

Exercise D

In each of the following questions, copy down the patterns replacing the question marks by the appropriate numbers or letters. n is any member of the set of counting numbers.

1 *Even numbers:*

the 1st even number is 2×1;

the 2nd even number is 2×2;

the 3rd even number is $? \times 3$;

the 4th even number is $? \times ?$;

....

the nth even number is   ?.

Sets and subsets

3 {2, 4, 8}.

No. (i) and (ii) Yes; the intersection set is always a subset of each of the intersecting sets.

4 Some possible pairs of subsets are:
{sons} and {daughters}; {infants} and {pensioners};
{those with blue eyes} and {those with brown eyes}.

3. FORMULAE

The section refers back to the use of letters to stand for elements in the set of counting numbers and statements involving counting numbers. It is really a revision of those occasions in *Book A* where such expressions were either used or hinted at. Where a pupil finds difficulty (as some do, particularly when stating the nth term of a sequence in terms of n), we feel that the best procedure might well be to move on through the course until a greater experience of the algebraic language makes these formulae clear.

Exercise D

1 2×3; 2×4; $2 \times n$.

Letters for numbers

2    $2\times3-1$; $2\times n-1$.

3   (i) 4; n;   (ii) $\dfrac{3\times4}{2}$; $\dfrac{n\times(n+1)}{2}$.

4

	V	F	E
Square-based pyramid	5	5	8
Cube	8	6	12
Octahedron	6	8	12

The equation is $V+F = E+2$.
Counting numbers; $V \geqslant 4$; $F \geqslant 4$; $E \geqslant 6$.

Formulae

2 *Odd numbers:*
   the 1st odd number is $2 \times 1 - 1$;
   the 2nd odd number is $2 \times 2 - 1$;
   the 3rd odd number is $? \times 3 - ?$;
   ....
   the nth odd number is   ?.

3 *Triangle numbers.*

(i) The difference between the 2nd and 1st triangle numbers is 2;
    The difference between the 3rd and 2nd triangle numbers is 3;
    The difference between the 4th and 3rd triangle numbers is ?;
    ....
    The difference between the nth and $(n-1)$th triangle numbers is ?.

(ii) By arranging triangle numbers in this manner,

• • •
• • •

we found that
   the 1st triangle number is $\dfrac{1 \times 2}{2}$;
   the 2nd triangle number is $\dfrac{2 \times 3}{2}$;
   the 3rd triangle number is $\dfrac{3 \times ?}{?}$;
   ....
   the nth triangle number is   ?.

4 *Polyhedra.* Sketch a tetrahedron, a square-based pyramid, a cube and an octahedron. If V stands for the number of vertices, F for the number of faces and E for the number of edges, complete the following table:

	V	F	E
Tetrahedron	4	4	6
Square-based pyramid	?	?	?
Cube	?	?	?
Octahedron	?	?	?

T 25

Letters for numbers

There is a connection between the numbers V, F and E. Write it down as an equation. V, F and E are members of what set of numbers? What is the least possible value that each can have?

5 *Angles*. (i) Two anticlockwise rotations follow each other and the result is a complete turn. If the first is a turns and the second is b turns, write down an expression for this result. Are a and b members of the set of counting numbers? What counting number is b less than?

(ii) If the rotations are measured in degrees, what counting number is b less than?

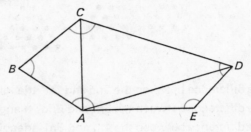

6 *Polygons*. (i) How many diagonals are there from any one vertex of

    (*a*) a 5-sided polygon;
    (*b*) a 6-sided polygon;
    (*c*) a 7-sided polygon;
    (*d*) an *n*-sided polygon?

(ii) How many diagonals are there altogether in a 4-sided; 5-sided; 6-sided; 7-sided and 8-sided polygon?

(iii)
How many more diagonals has a 5-sided than a 4-sided polygon?
How many more diagonals has a 6-sided than a 5-sided polygon?
How many more diagonals has a 7-sided than a 6-sided polygon?
How many more diagonals has an 8-sided than a 7-sided polygon?
How many more diagonals has an n-sided than an $(n-1)$ polygon?

Formulae

5   (i) $a+b = 1$. No. 1.   (ii) 360.

6   (i)   (a) 2;   (b) 3;   (c) 4;   (d) $n-3$.
    (ii)   2;   5;   9;   14;   20.
    (iii)   3;   4;   5;   6; $n-2$.

2. Tessellations

Tessellations are introduced in this chapter because they offer an opportunity to review in an interesting context much of the geometry which appeared in *Book A* and, at the same time, offer a starting point for future work on area, some aspects of motion geometry and the elementary study of pattern. They provide an interesting application of the angle properties of polygons and are closely linked with polyhedra—a plane tessellation is the special case of an infinite polyhedron. As with polyhedra, accuracy is again important for a satisfactory result and the desire to produce attractive designs will encourage pupils to acquire competence in geometrical construction. Tessellations help to develop a feeling for congruency and pattern. The latter interests many teachers and their pupils. In 1891 a Russian crystallographer named Fedorov showed that there are only 17 different ways in which a basic pattern can be repeated. Such repeating patterns are most commonly found on wall-papers, materials, and tiled floors. In the thirteenth century, the Arabs used all 17 ways when decorating the Alhambra in Granada, Spain.

For further reading on the 17 patterns see: Hilbert and Cohn-Vossen, *Geometry and the Imagination*; Wells, *The Third Dimension in Chemistry*; ane Coxeter, *Introduction to Geometry*.

Squared paper and isometric paper will be particularly helpful in this chapter.

1. PATTERN

The experiments in the Prelude are concerned with fitting together a finite number of tiles so that there are no empty spaces between them and so that no tile lies on top of another. In this section we consider how tiles can be arranged to form a repeating pattern which will fill a plane.

The polygons in Figure 1 have sides of equal length so that they can be used by pupils when answering Question 3 in Exercise B.

2. Tessellations

1. PATTERN

On tracing paper, mark the vertices of the regular polygons shown in Figure 1. Prick through the tracing on to card. Join the vertices with straight lines and cut out the polygons. You can then draw round your polygons to produce as many as you need. Do not lose them; you will need them to help you to answer several questions in this chapter.

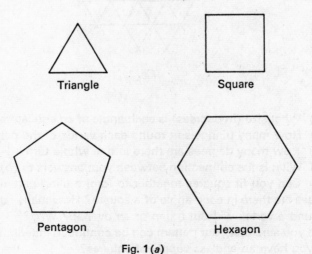

Fig. 1(a)

Tessellations

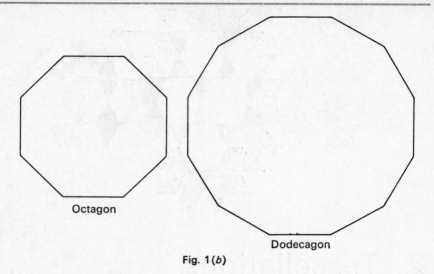

Fig. 1(b)

(a) Take your equilateral triangles and fit them together to form the tiling pattern shown in Figure 2.

Imagine that you have an endless supply of triangles: could you continue the pattern to fill the whole plane?

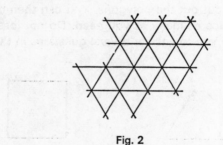

Fig. 2

(b) What size (in degrees) is each angle of an equilateral triangle?
(c) How many triangles fit round each vertex of the pattern?
(d) How many degrees are there in one whole turn?
(e) What is the connection between your answers to (b), (c) and (d)?
(f) Can you fit squares together to form a tiling pattern? How many degrees are there in each angle of a square? How many squares can you fit round a vertex without a gap or an overlap? Why?

Are you sure that your pattern can be continued indefinitely, supposing that you have an endless supply of squares?

(b) 60°.
(c) 6.
(d) 360°.
(e) 6 × 60° = 360°.
(f) Yes; graph paper or a chess board shows one possible way. 90°; 4; 4 × 90° = 360°.

Pupils will probably notice that there are other ways of covering a plane with squares or equilateral triangles (see Figure A).

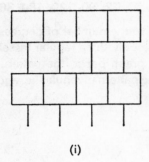
(i)
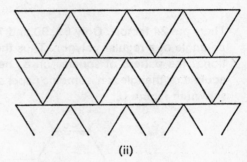
(ii)

Fig. A

The purpose of the last question is to make pupils aware that it must be possible to 'fill the plane' and one way of convincing others that this can be done is to produce a repeating pattern. Pupils tend to be somewhat haphazard in their early attempts to produce tessellations; shapes are added without a pattern emerging. Figure B shows a typical attempt with a 2 by 1 rectangle. It is hoped, however, that the experience gained from working through the experiments in the Prelude will have developed some awareness of the need for careful arrangement of tiles. In order to encourage an organized approach, several of the questions in Exercise A invite pupils to continue patterns which have already been partly drawn.

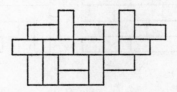

Fig. B

Tessellations

(*g*) 108°; no; 108 is not a factor of 360. Pupils should be allowed to experiment with their cut-out pentagons before being asked for an explanation. The practical discovery that these shapes do not fit is important.

(*h*) (i) Yes; 3×120° = 360°; (ii) no; (iii) no.

(*i*) 60° (see *Book A*, Chapter 7, Exercise C, Question 9). The set of factors of 360 is

{1, 2, 3, 4, 5, 6, 8, 9, 10, 12, 15, 18, 20, 24, 30,

36, 40, 45, 60, 72, 90, 120, 180, 360}.

There are 24 factors. Only 60, 90 and 120 give the number of degrees in an angle of a regular polygon. Thus there are only three regular tessellations: the pattern of squares on a sheet of graph paper, the pattern of equilateral triangles on isometric paper and the pattern of regular hexagons shown in Figure C.

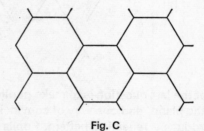

Fig. C

Exercise A

1 (*a*) See Figure D.

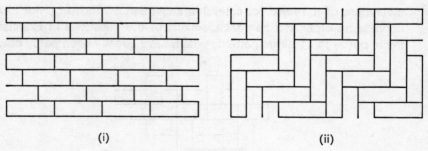

(i)   (ii)

Fig. D

Pattern

Patterns which fill a plane are called *tessellations*. Figure 2 shows a tessellation of equilateral triangles.

(*g*) What size (in degrees) is each angle of a regular pentagon? Can you form a tessellation of regular pentagons? If you are not sure, try fitting your regular pentagons together.

(*h*) Can you form a tessellation of:

        (i) regular hexagons;
        (ii) regular octagons;
        (iii) regular dodecagons?

Try to answer this question by thinking about the angles of these regular polygons. Check your decisions by fitting together your cut-out shapes.

(*i*) What is the smallest possible angle of a regular polygon?

List the set of factors of 360. How many factors are there? Compare your list with that of your neighbour. Which of your factors is the number of degrees in an angle of a regular polygon?

A tessellation of regular polygons, all of one kind and with their corners meeting at a point, is called a regular tessellation. How many regular tessellations do you think there are?

Exercise A

1 (*a*) Using squared paper to help you, copy and continue the patterns shown in Figure 3.

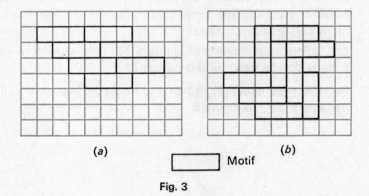

Fig. 3

(*b*) Draw a different tessellation of your own, using the same basic shape or motif.

(*c*) Where do you see tessellations of rectangles? Do the craftsmen who make them always use the same method? Look at the buildings which you pass on your way home.

Tessellations

2 (a) Copy and continue the patterns shown in Figure 4. Make sure that each pattern can be continued to fill the whole plane.

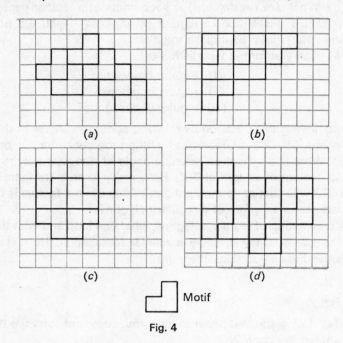

Fig. 4

(b) Colour your drawing of Figure 4 (d) so that shapes which are the same way up and the same way round are the same colour. How many colours do you need?

(c) Colour your drawing of Figure 4 (c) in the same way. How many colours do you need for this pattern?

3 (a) Using isometric graph paper to help you, copy and continue the patterns shown in Figure 5.

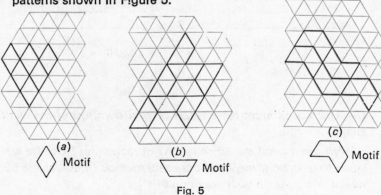

Fig. 5

Pattern

(*b*) Many tessellations are possible. Figure E shows two.

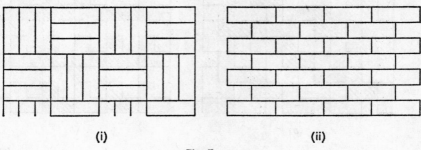

(i) (ii)
Fig. E

(*c*) Two of many brick patterns in actual use are shown in Figure F. Pupils could be asked to find as many as they can.

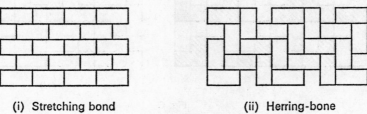

(i) Stretching bond   (ii) Herring-bone
Fig. F

It is well worthwhile to discuss with small groups of pupils such questions as:
Are all your shapes the same way up? Will your pattern look the same if it is turned upside down? Will it look the same if it is turned over? If a 'tile' is removed from your pattern, in how many ways can it be replaced? These and similar questions asked about a number of the patterns produced in this exercise will not only direct the pupils' attention away from the basic shape to the pattern as a whole, but also will prepare the way for work on symmetry in the next section and on motion geometry in later chapters. It may be that some teachers will wish to discuss transformations at this stage.

(*a*) See Figure G;   (*b*) two;   (*c*) four.

Tessellations

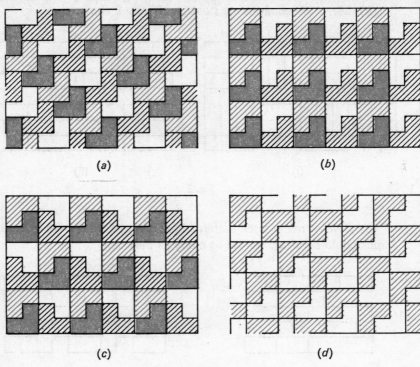

Fig. G

Pupils should be encouraged to colour their patterns in an organized manner. Time spent on colouring patterns is not wasted: indeed many pupils can see much more easily how to continue their patterns if colour is used. This is particularly true of a pattern like the one shown in Figure 4 (*d*).

This question suggests that shapes the same way up should have the same colour and this is perhaps the most helpful method for beginners, but it does nothing for the pattern shown in Figure 4 (*a*).

Another method, which makes each motif of the pattern easily distinguishable, is to colour patterns so that shapes with a common arc have different colours. It is possible to colour any plane pattern in this way using no more than four colours—a problem which is discussed in Chapter 11 of this book. Suitable colour schemes for (*a*) and (*b*) are shown in Figure G (*a*) and (*b*). These colour schemes emphasize the 'strip' technique which some pupils develop for forming tessellations. It is possible, however, to colour both of these patterns using only three colours. (*c*) and (*d*) need only two colours.

Pattern

3 Figure H shows three designs for each of the basic shapes. There are others.

Fig. H

Tessellations

4 Figure I shows two possible tessellations. Only three colours are necessary but most pupils will discover these patterns more easily if shapes orientated in the same way have the same colour. In this case, 4 colours would be used for Figure I (i) and 6 for Figure I (ii).

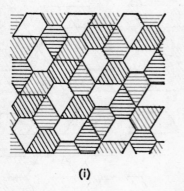

(i)   (ii)

Fig. I

5 Figure J shows one possible pattern. A square and an equilateral triangle.

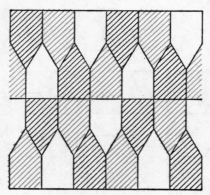

Fig. J

6 (*b*) Three colours are needed.

Pattern

(b) Draw at least two different tessellations for each of the three motifs.

4 The motif shown in Figure 6 is formed from a regular hexagon and an equilateral triangle. Use this motif to draw two different tessellations.

Fig. 6

5 Sketch a tessellation using the motif shown in Figure 7. Which two simple shapes are used to build this motif?

Fig. 7

6 (a) On a sheet of plain paper draw an accurate tessellation of regular hexagons. Use the method suggested by Figure 8.
   Are you satisfied with the appearance of your drawing? If not, do it again.

   (b) Using as few colours as possible, colour your drawing so that no hexagon touches another hexagon of the same colour.

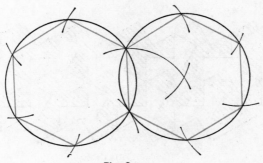

Fig. 8

Tessellations

7   Equipment: tracing paper, thin card, scissors.

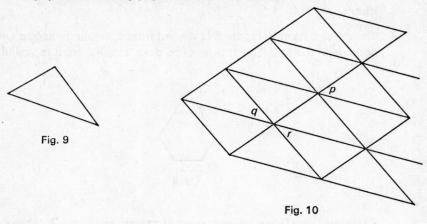

Fig. 9

Fig. 10

(*a*) Trace the triangle shown in Figure 9 on to card. Cut out triangles the same shape and the same size. Try to get as many as possible from your card. Paste about half your triangles into your exercise book to form the tessellation shown in Figure 10.

(*b*) Mark all the angles in your figure which are equal to the one marked with the letter *p*.

(*c*) What can you say about the angles marked with the letters *q* and *r*?

(*d*) Can you form any different tessellations with the remainder of your triangles? When you have a new pattern, record it either by pasting triangles in the correct positions in your exercise book or by pricking through the vertices of a tracing of your triangle on to a sheet of plain paper.

8   Figure 11 shows how a regular tessellation can be used to form more intricate designs. Try some for yourself.

Fig. 11 (*a*)

T40

7 (a) and (b). This question illustrates the traditional properties of parallels such as alternate and corresponding angles. Accuracy is important for a satisfactory result.
   (c) The angles marked with the letters q and r are equal.
   (d) It is necessary to turn the triangle over in order to obtain the pattern shown in Figure K.

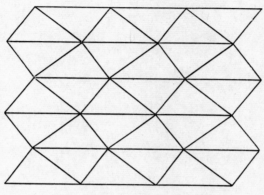

Fig. K

8 The design in Figure 11 (a) is based on the regular tessellation of equilateral triangles and that in Figure 11 (b) on the regular tessellation of squares.
   Some remarkable tessellations can be found in *The Graphic Work of M.C. Escher*.

Tessellations

2. SYMMETRY

(a) $\frac{1}{3}$ turn (120°), $\frac{2}{3}$ turn (240°), 1 turn (360°); order 3.

Pattern

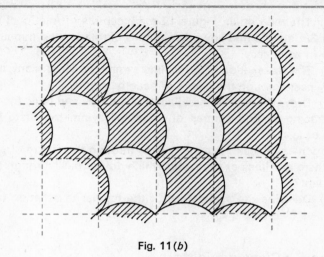

Fig. 11(*b*)

2. SYMMETRY

Figure 12 shows part of a tessellation of regular hexagons.

(*a*) Copy this tessellation on to isometric graph paper and make a second copy on tracing paper.

Prick a compass point through the vertex marked *A* and rotate the tracing paper until the two patterns are again superimposed.

Through what angle have you rotated the pattern? Is there more than one possible angle?

What is the order of rotational symmetry about the vertex *A*? (Imagine that the tessellation is extended to fill the whole plane.)

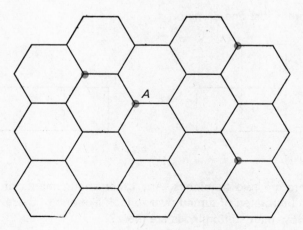

Fig. 12

Tessellations

(b) The red dots in Figure 12 mark centres of rotational symmetry of order 3. Can you find any other centres of rotational symmetry? State the order in each case.

(c) The tessellation also has line symmetry. How many lines of symmetry pass through the centre of each hexagon?

(d) Draw the regular tessellation of squares and trace the pattern. Use the tracing to find centres of rotational symmetry. State the order in each case.

How many lines of symmetry pass through the centre of each square? Are there any lines of symmetry which do not pass through the centre of a square?

(e) Describe the symmetries of the regular tessellation of equilateral triangles (see Figure 2).

Exercise B (Class projects)

1 Shapes made from squares joined edge to edge are sometimes called polyominoes.

        Domino :   made from two squares.
        Tromino :   made from three squares.
        Tetromino : made from four squares.
        Pentomino : made from five squares.

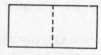

Fig. 13

There is just one domino.

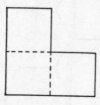 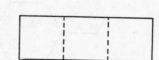

Fig. 14

There are two trominoes. Any other arrangements of three squares can be rotated or turned over to look like one of these.

How many tetrominoes are there?
How many pentominoes?

Symmetry

(b) Vertices of hexagons, order 3.
Middle points of sides of hexagons, order 2.
Centres of hexagons, order 6.
(c) 6.
(d) Vertices of squares, order 4.
Middle points of sides of squares, order 2.
Centres of squares, order 4.
Four lines of symmetry pass through the centre of each square. The straight lines formed by the edges of the squares are also lines of symmetry.
(e) Vertices of triangles, order 6.
Middle points of sides of triangles, order 2.
Centres of triangles, order 3.
Six lines of symmetry pass through each vertex.

Exercise B

Class projects

1 The five tetrominoes are shown in Figure L, the twelve pentominoes and the thirty-five hexominoes can be found in the *Teacher's Guide to Book A* on pages T263 and T260 (all mirror images are excluded).

Fig. L. The five tetrominoes.

This is the first time that pupils have been asked for *all* the shapes made from a fixed number of squares. One way of ensuring that all the tetrominoes, for example, have been found is to take the trominoes and fit a monomino in all possible positions in turn (see Figure M) and then reject repetitions. It is important that this method should not be suggested to pupils unless they have both felt the need for a method and had an opportunity to find a method of their own.

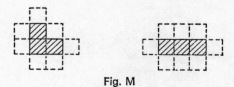

Fig. M

Tessellations

A shape made from seven squares which cannot be tessellated is shown in Figure N.

Fig. N

2 Once again pupils are asked to exhaust a situation. The shapes are shown in Figure O. Tessellations can be formed from all of them.

Tessellations can also be formed from all twelve of the hexiamonds but not from all of the twenty-four heptiamonds.

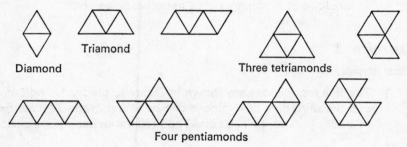

Fig. O

3 A semi-regular tessellation is composed of regular polygons of two or more kinds so that the arrangement of corners is the same at every vertex. The tessellation is therefore specified by giving in order the polygons which surround each vertex. Figure 15 (a) shows the 3.6.3.6 pattern.

$3^4.6$ means that four triangles and one hexagon surround each vertex. If a notation for this pattern is used in the classroom, then it is less confusing for some pupils if the pattern is called 3.3.3.3.6 rather than $3^4.6$.

Symmetry

Draw sketches to show that tessellations can be made from all the polyominoes with one, two, three, four or five squares.

The shapes made from 6 squares are called hexominoes. Tessellations can be formed from all of them. Draw some of these tessellations and colour them so that a shape does not have an edge in common with a shape of the same colour (see Figure 11).

Can you find a shape made from 7 squares with which you cannot form a tessellation?

2 Shapes made from equilateral triangles joined edge to edge are sometimes called polyiamonds.

      Diamond: made from two equilateral triangles.
      Triamond: made from three equilateral triangles.
      Tetriamond: made from four equilateral triangles.
      Pentiamond: made from five equilateral triangles.

How many triamonds, tetriamonds and pentiamonds are there? Can you form tessellations from all of these?

3 There are only 3 regular tessellations but other patterns can be formed from regular polygons when more than one kind are used. Use your cut-out polygons to find as many of these patterns as you can. Figure 15 shows two examples.

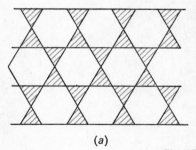 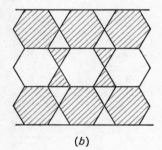

(a)   (b)

Fig. 15

When you find a pattern, record it. One attractive way of doing this is to cut the shapes from coloured gummed paper, stick them on to plain paper and display the finished patterns in the classroom.

When the shapes surrounding every vertex are in the same order, the tessellation is called semi-regular. Each vertex in Figure 15(a) is surrounded by hexagon, triangle, hexagon, triangle, in that order. In Figure 15(b), some vertices are like those in Figure 15(a); others are surrounded by hexagon, triangle, triangle, hexagon. The first pattern is semi-regular; the second is not. Sort your patterns into two sets:

Tessellations

those that are semi-regular and those that are not. There are eight semi-regular tessellations. Have you found all of them?

4 Make a collection of all the polyhedra which you have made. Which of them can be arranged to fill space?

5 Describe the symmetries of each of the eight semi-regular tessellations.

6 Design some tessellations with rotational symmetry of order 4. If possible, design a tessellation with rotational symmetry of order 4 only (that is, with no lines of symmetry).

Summary

Patterns which completely 'fill' the plane are called *tessellations*. A tessellation can be built from one basic unit, for example, a parallelogram. A tessellation can also be built from two or more basic units, for example, a square and a regular octagon.

A *regular tessellation* is a tessellation of regular polygons, all of one kind, and such that every vertex of the pattern is exactly like every other vertex. There are three regular tessellations: the pattern of squares on a sheet of graph paper, the pattern of equilateral triangles on isometric paper and the pattern of regular hexagons shown in Figure 12.

Symmetry

The eight semi-regular plane tessellations are: $3^3.4^2$, $3^2.4.3.4$, $3.6.3.6$, $3^4.6$, $3.4.6.4$, 3.12^2, 4.8^2, and $4.6.12$ (see Figure P).

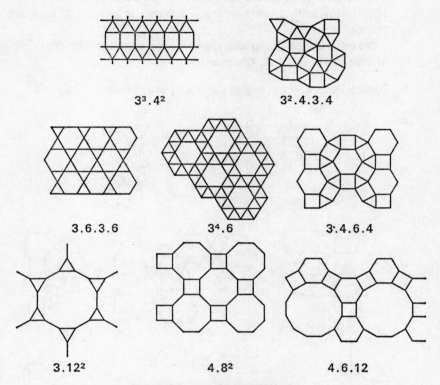

Fig. P. The eight semi-regular tessellations.

The three regular tessellations are: 3^6, 4^4, 6^3.

An unlimited number of tessellations can be formed from regular polygons if it is not required that all vertices should be the same. Some of these patterns are extremely attractive (see Steinhaus, *Mathematical Snapshots*).

Tessellations

4 Space-filling arrangements can be made with
  (i) cubes;
  (ii) regular octahedra and regular tetrahedra together;
  (iii) prisms with cross-sections which can be tessellated, for example, triangular prisms.

  No other space-filling arrangements can be made with the polyhedra mentioned in *Book A*, Chapter 10.

5 Typical lines and centres of symmetry are shown in Figure Q.

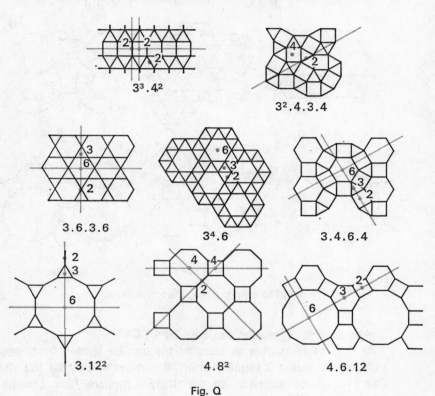

Fig. Q

6 Figure R shows a tessellation with rotational symmetry of order 4 and with no lines of symmetry. There are, of course, also centres of rotational symmetry of order 2 since the design is based on the regular tessellation 4^4.

Fig. R

Tessellations

WORK CARDS

The following cards contain suggestions for further work on tessellations and polyhedra.

4 pupils   No. 1   Hexagonal animals

How many different shapes can you make with 1, 2, 3, 4, ... regular hexagons joined edge to edge?

Which of your shapes can be used to form tessellations?

4 pupils   No. 2   Truncated octahedra

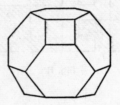

Fig. S

Find out how you can make a truncated octahedron—in which the vertices of a regular octahedron are cut off to form a polyhedron with 8 regular hexagonal faces and 6 square faces.

Can truncated octahedra be used to fill space?

4 pupils   No. 3   Semi-regular polyhedra

The faces of a semi-regular polyhedron are regular polygons of two or more kinds and every vertex is exactly like every other vertex. Use cut-out regular polygons to discover some semi-regular polyhedra. Design suitable nets and make models.

When you have found as many as you can, look at *Mathematical Models* by Cundy and Rollett to see if you can find some others.

3. Decimals

In this chapter, the introduction to decimals is placed in a practical context, that of measurement. As soon as pupils start making their own measurements and comparing their answers, the practical difficulties (and inevitable inaccuracies) become apparent and so do the needs for units and parts of units and a method of notation.

The central development of this chapter is as follows: an introduction to the problem of measurement; the need for a standard unit and the division and sub-division of that unit; the use of decimal notation to help us conveniently state this division of units. As the metric system is now to be used in our practical, everyday measurements, we have, for the first time in this country, units which accord with our number system; metric units of length are introduced in this chapter. The final section forms an introduction to rounding off and significance.

Decimals

1. MEASUREMENT

The idea that measurement involves a comparison with a standard unit is introduced early in the chapter. From this should follow the realization that an exact statement of the length of any line in terms of physical measurement cannot be made. This is not always easily accepted by pupils. The distinction between counting and measuring is made in Section 1.2. (Question 1 of Exercise A provides a basis for discussion on these points.)

Though the need for some standard unit of length is established in Section 1.1, the metre is not introduced until Section 1.3. This is done deliberately because, for some years after the publication of this book, some pupils will probably be more familiar with Imperial than with metric units and the authors wish to avoid introducing the idea of a necessary standard unit at the same time as the unfamiliar notation and experience of a new unit. The use of other metric units, for example, weight, will occur from time to time in the text and in exercises.

Pupils should be encouraged to notice that all measurements involve the choice of a suitable unit and an appreciation of the degree of accuracy for which they should aim. This appreciation will develop with experience: in this section pupils should be encouraged to discuss whether certain answers can sensibly be expected with the instruments available, or whether better instruments would be required to achieve them.

It is important that pupils realize that a measurement given to 2 decimal places is not necessarily more correct than one given to the nearest whole number. A large number of decimal places may be quite meaningless if the measurements have obvious limitations of accuracy. In all cases the degree of accuracy, both of measurements and calculations, should be clearly stated.

3. Decimals

1. MEASUREMENT
1.1 Units of length

Take *any* pencil and measure to see how many 'pencil-lengths' wide your desk is. Compare results. Did you find that everyone having the same sized desk as you, had the same result? If not, why not?

Can you think of a better way of comparing or measuring the widths of your desks?

How many centimetres wide is your desk? Did you get a whole number of centimetres for your answer? Compare results with anyone having the same sized desk as yourself. Even now, when everybody is using the *same unit of length* to measure with, you probably do not agree entirely as to the width of your desks. Why not?

1.2 Fractions of units

You have probably already found the need for fractions of a unit when trying to give fairly accurate measurements. What fraction of a unit did you use?

Figure 1 shows the measurement of the line segment AB. Each unit is divided into tenths. We could say that the length of AB is

   (*a*) a bit more than 2 units;

or, to be more precise,

   (*b*) just over $2\frac{3}{10}$ units.

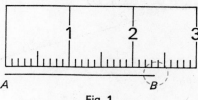

Fig. 1

Decimals

Alternatively, we could say that the length of *AB* is

    (*a*) 2 units (to the nearest unit);

or, to be more precise,

    (*b*) $2\frac{3}{10}$ units (to the nearest tenth of a unit).

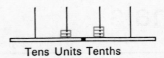

Fig. 2

Figure 2 shows the number $2\frac{3}{10}$ represented on an abacus. We could write this as $2\cdot3_{\text{ten}}$, using a fraction point. What would one disc on the unlabelled spike represent?

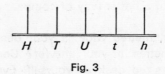

Fig. 3

Figure 3 shows an abacus designed for use in decimal arithmetic. What do the letters *H, T, U, t* and *h* stand for?

When we are working in base ten, we usually refer to the 'fraction point' as the 'decimal point'. We usually leave out the word ten, writing $2\cdot3_{\text{ten}}$ as $2\cdot3$.

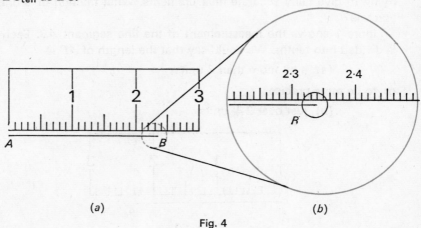

(*a*)   (*b*)

Fig. 4

Measurement

1.2 Fractions of units

Parts of a unit are obviously necessary if any attempt at precise measurement is to be made, particularly when a large unit of measure is involved. With metrication, tenths and hundredths are the obvious ones to use.

The now familiar abaci of Figures 2 and 3 should help with the explanation of Figure 4.

Decimals

Exercise A

1 (*a*), (*d*), (*f*), and (*h*) are examples of counting and can therefore be given an exact measure.
   In (*b*), (*c*), (*e*), (*g*) and (*i*), the practical problems of the measurements could be discussed.

2 (*a*) 5; (*b*) 0·03;
  (*c*) 300; (*d*) 2·1;
  (*e*) 6·05; (*f*) 700·6.

3 (*a*) 4 tens; (*b*) 4 units; (*c*) 4 tenths;
  (*d*) 4 thousands; (*e*) 4 hundredths.

Measurement

Figure 4 shows us how to give the length of the line segment *AB* more accurately than before. Figure 4(*b*) is an enlargement of the end of the segment near *B*. Each tenth of a unit has been divided into ten smaller equal parts (hundredths of a unit). We can now say that the length of *AB* is

(*c*) 2·34 units (to the nearest hundredth of a unit).

Can we say that *AB* is *exactly* 2·34 units long? Suppose you were to look at the black circled part, through a microscope. You might find that *AB* was just a little more than or just a little less than 2·34 units.

When we measure any length, we *compare* it with a fixed *unit*. It is because we are only *comparing* and not *counting* that we cannot talk about an *exact* length. Even if we divided the unit into smaller parts, we could not be certain that we had made an exact comparison.

Whenever we state a measure, which is a number, we must follow it with the unit we have used for comparison.

Exercise A

Which of the following activities involve measurement and which require only counting? In which is it possible to give an exact answer?

1. (*a*) Finding the number of pupils in your school;
   (*b*) finding the distance between your home and school;
   (*c*) finding your own weight;
   (*d*) finding the number of steps to the top of St Paul's Cathedral;
   (*e*) finding the temperature of a saucepan of milk;
   (*f*) finding the number of peas in a packet of frozen peas;
   (*g*) finding the time you take to walk one kilometre;
   (*h*) finding the number of words on a page of a book;
   (*i*) finding your age at mid-day on January 1st of this year.

2. Write the following in number symbols using the decimal point when necessary. (Remember that when we write whole numbers we do not need to put in the decimal point, but its position *would be* immediately after the digit which represents the units.)

   (*a*) Five units;   (*b*) three hundredths;
   (*c*) three hundreds;   (*d*) two and one tenth;
   (*e*) six and five hundredths;   (*f*) seven hundreds and six tenths.

3. What does the 4 stand for in each of these numbers?
   (*a*) 41·6;   (*b*) 1004;   (*c*) 6·4;
   (*d*) 4321;   (*e*) 0·04.

Decimals

4   Draw an abacus like the one in Figure 3 to represent each of the following 'decimal numbers':

(a) $7\frac{3}{10}$;   (b) $24\frac{7}{10}$;   (c) $123\frac{4}{10}$;
(d) 6·8;   (e) 314·25;   (f) 62·08;
(g) $13\frac{23}{100}$;   (h) $106\frac{37}{100}$;   (i) 0·32.

5   Explain the difference between the following numbers. What does the nought tell you in each case?

(a) 62·01 and 62·1;   (b) 38 and 380;
(c) 47·16 and 47·106;   (d) 13 and 103.

6

?	Hundreds	Tens	Units	Tenths	Hundredths	?
5	2	0	0			
	5	2	0			
		5	2			

(a) What column headings would you give to the unlabelled columns?

(b) By completing the table, add three more numbers to the sequence 5200, 520, 52, ..., ..., ... What is the connection between each number and the next number to the right?

Draw six abaci, one under the other, and show one of these numbers on each. What is the connection between the discs on the spikes and the numbers in the columns?

Add three more numbers to the following sequences:

(c) 93 000, 9300, 930, ..., ..., ...;

(d) 0·019, 0·19, 1·9, ..., ..., ...

What is the connection between each number and the next number to the right in (c) and (d)?

2. STANDARD UNITS OF LENGTH

The standard unit of length is the metre (m). Other units are often used but they are either multiples or decimal fractions of the metre.

For smaller objects, the centimetre (cm) is used, where 100 cm = 1 m. (There is another small unit, the millimetre (mm), where 1000 mm = 1 m.) For longer distances, the kilometre (km) is used, where 1 km = 1000 m.

T60

Measurement

4

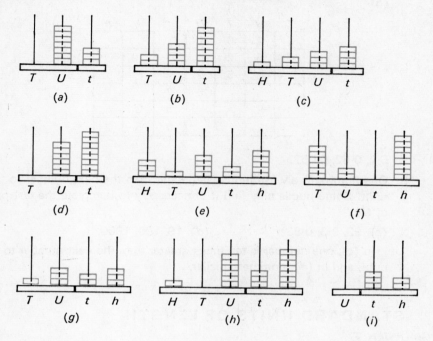

5 That:
   (a) there are no tenths;   (b) there are no units;
   (c) there are no hundredths; (d) there are no tens.

Decimals

6 (a) Thousands and thousandths.

   (b)

T	H	T	U	t	h	th
5	2	0	0			
	5	2	0			
		5	2			
			5	2		
				5	2	
				0	5	2

5·2, 0·52, 0·052.

One number is always ten times greater than the next number to the right. Some pupils may find it unnecessary to complete the last part of (b).

(c) 93, 9·3, 0·93;   (d) 19, 190, 1900.

In (c), one number is ten times greater than the next number to its right, and in (d) ten times smaller.

2. STANDARD UNITS OF LENGTH

Exercise B

1 (a) m;   (b) m;   (c) m or cm;   (d) cm or mm.

   The making of these measurements can form a useful basis for discussion and the teacher might like to add examples of his own.

2 (a) m;   (b) km;   (c) cm or mm;
   (d) mm;   (e) km.

3 (a) 3 cm, 2·6 cm;   (b) 4 cm, 4·2 cm;
   (c) 3 cm, 3·3 cm;   (d) 2 cm, 1·6 cm;
   (e) 1 cm, 1·0 cm;   (f) 3 cm, 2·6 cm.

   In the second part of each question, measurements to 1 decimal place either way should probably be accepted.

   Pupils should generally be able to estimate to the nearest centimetre, lengths of less than 5 or 6 cm by eye.

Standard units of length

$$1 \text{ km} = 1000 \text{ m},$$
$$1 \text{ cm} = \frac{1}{100} \text{ m},$$
$$1 \text{ mm} = \frac{1}{1000} \text{ m}.$$

Each unit can be expressed as a multiple or a decimal of the other. For example:

1 m	= 100 cm;	100 cm	= 1 m;
4 m	= 400 cm;	1 cm	= 0·01 m;
0·5 m	= 50 cm;	18 cm	= 0·18 m;
1·73 m	= 173 cm;	160 cm	= 1·60 m;
1 km	= 1000 m;	1000 m	= 1 km;
2 km	= 2000 m;	1 m	= 0·001 km;
1·6 km	= 1600 m;	97 m	= 0·097 km;
0·05 km	= 50 m;	365 m	= 0·365 km.

Exercise B

1 With what units of length would you choose to measure:
   (a) the width of your classroom;
   (b) the height of your classroom;
   (c) your own height;
   (d) the thickness of this book?
   Now carry out these measurements. What practical difficulties did you find?

2 With what units of length would you choose to measure the following:
   (a) the length of your garden;
   (b) the distance from London to Glasgow;
   (c) the length of a match;
   (d) the thickness of a sheet of paper;
   (e) the distance between Earth and Mars?

3 Measure the lengths of these line segments. Give each length:
   (i) to the nearest centimetre;
   (ii) to the nearest tenth of a centimetre.
   (a) ──────────────   (b) ──────────────────────
   (c) ────────────────   (d) ─────────
   (e) ──────   (f) ──────────────

   Could you *by eye* estimate each length correct to the nearest centimetre?

Decimals

4 Draw, as accurately as you can, line segments of the following lengths:

(a) 4 cm; (b) 2·3 cm; (c) 1·8 cm;
(d) 0·7 cm; (e) 10·5 cm; (f) 15·8 cm;
(g) 13·2 cm; (h) 20·1 cm.

Will any of your line segments be exactly the length mentioned?

5 Which of the following statements would you criticize and why?

(a) I am exactly 1·52 m tall.

(b) There are exactly 100 cm in 1 m.

(c) I made this dress in 8 h 35 min.

(d) I measured this line segment with my ruler and it is exactly 3·8 cm long.

(e) There are exactly 1000 m in 1 km.

3. ADDITION AND SUBTRACTION SHOWN ON THE ABACUS

Addition and subtraction are easy when you are working with whole numbers, and they are just as easy when you are working with numbers which involve a decimal point.

Example 1

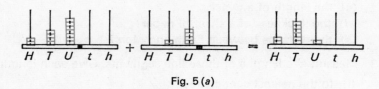

Fig. 5 (a)

This figure shows the addition

$$247 + 14 = 261 \quad \text{or} \quad \begin{array}{r} 247 \\ +14 \\ \hline 261 \end{array}$$

Addition and subtraction

5   (a), (d). It is not possible to measure any distance *exactly*.
(c) Time given to the nearest quarter or half hour would be more meaningful.
(b), (e). These are true by definition.

3. ADDITION AND SUBTRACTION SHOWN ON THE ABACUS

Many pupils cannot see the connection between work with integers and that with decimals, but the use of pairs of drawings like those in Figures 5 and 6 should help them.

Decimals

Example 1
   Only the position of the digits relative to the decimal point has changed.
   (i) 2610,   (ii) 0·261.

Example 2
   0·31 − 0·14 = 0·17.

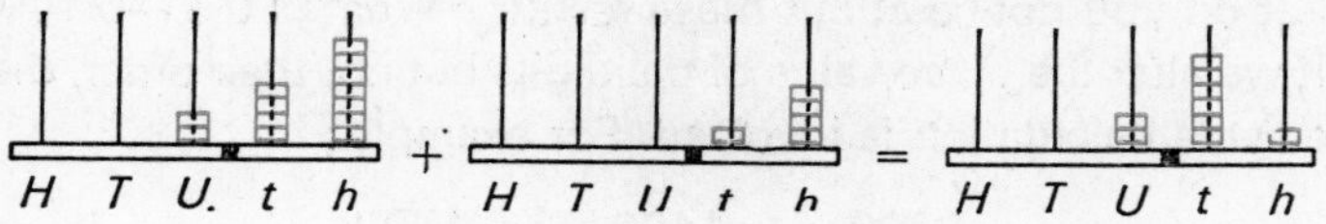

Addition and subtraction

[Fig. 5(b) abacus diagram]

Fig. 5(b)

This figure shows the addition

$$2 \cdot 47 + 0 \cdot 14 = 2 \cdot 61 \quad \text{or} \quad \begin{array}{r} 2 \cdot 47 \\ + 0 \cdot 14 \\ \hline 2 \cdot 61 \end{array}$$

What do you notice about the two results?
What are: (i) 2470+140, (ii) 0·247+0·014?

Example 2

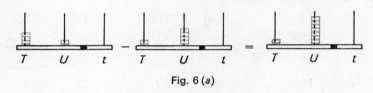

Fig. 6(a)

This figure shows the subtraction

$$31 - 14 = 17 \quad \text{or} \quad \begin{array}{r} 31 \\ -14 \\ \hline 17 \end{array}$$

[Fig. 6(b) abacus diagram]

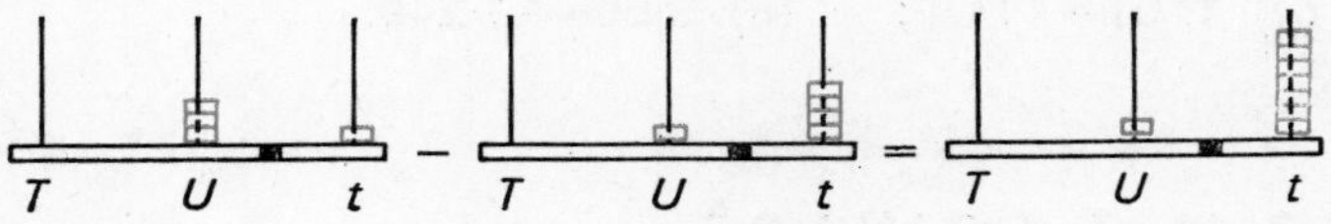

Fig. 6(b)

This figure shows the subtraction

$$3 \cdot 1 - 1 \cdot 4 = 1 \cdot 7 \quad \text{or} \quad \begin{array}{r} 3 \cdot 1 \\ -1 \cdot 4 \\ \hline 1 \cdot 7 \end{array}$$

T 67

Decimals

What do you notice about these results? What is 0·31−0·14?

If we alter the place value of the digits but not their order, then basically the same calculation is involved. For example:

$$3100 \quad -1400 \quad = 1700$$
$$310 \quad - 140 \quad = 170$$
$$31 \quad - 14 \quad = 17$$
$$3·1 \quad - 1·4 \quad = 1·7$$
$$0·31- \quad 0·14 = \quad 0·17$$

Exercise C

1   Draw a pair of diagrams to show the following calculations:
   (a) 36·1+27·3,   (b) 623−111,
       3·61+2·73;   6·23−1·11.

   For each pair of calculations, what do you notice?

2   Work out the following:

   (a) + 4·91   (b) + 23·04   (c) + 104·3
            3·68   8·19   0·12

   (d) − 16·8   (e) − 3·87   (f) − 2·11
            2·9   0·08   1·95

3   Set out the following in the way shown in Question 2, then work them out.
   (a) 2·9+0·11;   (b) 3·64−0·91;   (c) 20·07+104;
   (d) 34·36−17·82;   (e) 81·6−1·92;   (f) 10·1−0·08;
   (g) 1101−11·01;   (h) 3·508−1·249.

4. DECIMAL COINAGE

Addition and subtraction

Exercise C

1 (a)

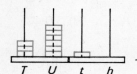 + 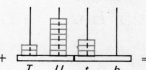 =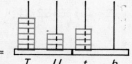

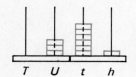 + 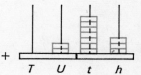 =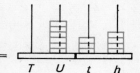

(b) Similar pair of diagrams showing

623 − 111 = 512 and 6·23 − 1·11 = 5·12.

2 (a) 8·59;   (b) 31·23;   (c) 104·42;
  (d) 13·9;   (e) 3·79;   (f) 0·16.

3 (a) 3·01;   (b) 2·73;   (c) 124·07;
  (d) 16·54; (e) 79·68;   (f) 10·02;
  (g) 1089·99; (h) 2·259.

Decimals

4. DECIMAL COINAGE
Exercise D
1 (a) 109·1;   (b) 32·1;   (c) 122·7;   (d) 0·735.
2 (a) £152·31;   (b) £1·19;   (c) £1·75;   (d) £0·43.

Decimal coinage

We have just seen that it is not difficult to add and subtract decimal numbers. This is one of the reasons why most countries use a decimal coinage system. Let us look at the system that has the 'pound' as the basic unit. This is divided into 100 smaller units called 'new pence' (p) so that:

$$100p = £1$$

and

$$1p = £\tfrac{1}{100} = £0·01.$$

Any number of new pence can be expressed as a decimal of a pound. For example:

$$2p = £0·02;$$
$$3p = £0·03;$$
$$10p = £0·10;$$
$$25p = £0·25;$$
$$70p = £0·70;$$
$$125p = £1·25.$$

Notice that when *less than* 100p is expressed in pounds, a zero is always written before the decimal point. If we did not do this, we might confuse £·25 with £25.

Notice also that when stating parts of a pound, a digit is shown in *both* of the first two decimal places. £2·50 should never be written as £2·5, for this might be taken as £2 with only 5p, instead of 50p.

Exercise D

Work out the following:

1. (a) 17·1 + 92   (b) 83·2 − 51·1   (c) 146·1 − 23·4   (d) 0·761 − 0·026

2. (a) £ 70·93 + 81·38   (b) £ 2·67 − 1·48   (c) £ 3·42 − 1·67   (d) £ 1·31 − 0·88

Decimals

3 2·47 + 3·88 − 1·72.

4 £8·17 + £80·19 − £4·29.

5 Use your answers to Question 1 to *write down* the answers to the following:

(a) $+\begin{array}{r} 1\cdot71 \\ 9\cdot2 \end{array}$ (b) $-\begin{array}{r} 832 \\ 511 \end{array}$ (c) $-\begin{array}{r} 1\cdot461 \\ 0\cdot234 \end{array}$ (d) $-\begin{array}{r} 761 \\ 26 \end{array}$

6 The United States of America has a decimal coinage system based on the 'dollar' ($) which is divided into 100 smaller units called cents (c). European countries have similar systems based on a 'franc', 'mark', etc., which are divided into 100 smaller units.

Work out the following:

(a) $+\begin{array}{r} \$ \\ 108\cdot70 \\ 92\cdot83 \end{array}$ (b) $-\begin{array}{r} \$ \\ 23\cdot21 \\ 19\cdot53 \end{array}$ (c) $+\begin{array}{r} \text{Fr.} \\ 1\cdot66 \\ 99\cdot71 \end{array}$ (d) $-\begin{array}{r} \text{Fr.} \\ 32\cdot91 \\ 26\cdot99 \end{array}$

5. ROUNDING OFF AND SIGNIFICANCE
5.1 Rounding off

Is the length of this page 23·43 cm to the nearest hundredth of a centimetre? Can you measure as accurately as this with your ruler?

An ordinary ruler is marked in tenths of centimetres or, sometimes, only in whole centimetres. In such cases, we can check that the page length is 23·4 cm to the nearest tenth or 23 cm to the nearest whole centimetre.

Thinking of 23·43 cm to the nearest hundredth, as 23·4 cm to the nearest tenth, or as 23 cm to the nearest unit, is called 'rounding off'.

A sum of £2 467 983 spent on some national scheme, might well be rounded off to the nearest million pounds as £2 000 000, just for convenient reference. What is this sum of money rounded off to the nearest (i) half million; (ii) thousand pounds?

The number of people in a crowded stadium one day, is 29 381. To make this easy to remember, it might be rounded off to the nearest thousand as 29 000. What would it be to the nearest ten thousand?

What is the length of the line segment *AB* in Figure 7 (i) to the nearest tenth of a unit, (ii) to the nearest unit?

Decimal coinage

3   4·63.

4   £84·07.

5   (a) 10·91;   (b) 321;   (c) 1·227;   (d) 735.

6   (a) $201·53;   (b) $3·68;   (c) 101·37 fr;   (d) 5·92 fr.

5. ROUNDING OFF AND SIGNIFICANCE

This is a preliminary discussion on these subjects. Rounding off is something that needs constant practice and some pupils find it rather difficult, so the work has been divided into two parts, this section and a later chapter on computation and applications of the slide rule. It is not expected that pupils should be experts at hard examples of rounding off by the end of this section, but they should have developed a general awareness of significance. There will be ample opportunity for further discussion and revision in later books. The problem of limits of accuracy will be considered later in the course.

5.1 Rounding off

(i) £2 500 000;   (ii) £2 468 000.

Pupils may find difficulty in picking out the number of significant digits necessary to give the sum of money correct to the nearest thousand pounds. If so, ask the question 'Is £2 467 983 nearer to £2 467 000 or £2 468 000?'

29 381 = 30 000 (to the nearest ten thousand).

Decimals

5.2 Significant figures

£3780 = £4000 (to 1 S.F.)
 = £3800 (to 2 S.F.).

Exercise E

1. (a) 7·1 m; (b) 10·4 m; (c) 22·2 m;
   (d) 103·7 cm; (e) 8·6 cm; (f) 0·9 cm;
   (g) 0·9 km; (h) 1·0 km; (i) 8·5 km.

Note in particular (h) and (i). More examples of this kind might be added.

Rounding off and significance

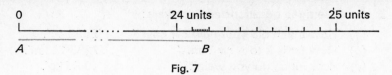

Fig. 7

When rounding off to the nearest tenth, 24·16 becomes 24·2, as the '6 hundredths' indicate that we are nearer 2 tenths than 1 tenth. In rounding off to the nearest unit, we have 24, as the '1 tenth' indicates that we are nearer 4 units than 5 units.

It is customary to round off a number such as 24·1<u>5</u>, 'upwards' as 24·2.

5.2 Significant figures

(a) Numbers are rounded off for various reasons: because of inaccurate instruments, for ease of reference and to make them easier to remember. We also round off so that we can concentrate upon the most important or *significant* digits. In the numbers

190·3 and 0·01903,

which digit is the most significant? Does this depend upon the size of the digit or upon its place value?

The most significant digits are always those whose *place value* is greatest, that is, those on the left. The '1' is more significant than the '9'. The '9' is more significant than the '3'.

The number of significant digits or 'figures' we choose, depends upon the circumstances. A distance of 17·5 km to a long distance runner in training might be rounded off to 18 km by a man who is walking for pleasure or to 20 km by a motorist. The runner is interested in 3 significant figures (we write this as '3 S.F.'), the walker in 2 S.F. and the motorist in 1 S.F.

A house is for sale at £3780. Write this number rounded off to 1 S.F. and again to 2 S.F. A rich man might well think in thousands of pounds. To him there is only one significant figure, the '4' in £4000. A poorer man might worry about the hundreds of pounds. To him there are two significant figures, the '3' and '8' in £3800.

Exercise E

1 Write down the following lengths rounded off to the nearest tenth of a unit:

(a) 7·13 m; (b) 10·39 m; (c) 22·22 m;
(d) 103·66 cm; (e) 8·55 cm; (f) 0·85 cm;
(g) 0·93 km; (h) 0·99 km; (i) 8·50 km.

Decimals

2   Here are three questions and answers:
   (a) 'How much did your new car cost?'   '£693·47'.
   (b) 'How far is it to New York?'   '7283·14 km'.
   (c) 'How much do you weigh?'   '49·37 kg'.

   In each case, say why the answer is ridiculous and give a sensible answer.
   To how many significant figures did you give your answers? Are you necessarily going to agree with your friends?

3   Write down the following numbers rounded off to (i) 1 S.F. and (ii) 2 S.F.:
   (a) 142;   (b) 177;   (c) 949;   (d) £63·15;
   (e) 12·14 km;   (f) 23·2 m;   (g) 15 km;   (h) £301;
   (i) £7004;   (j) 1040 km;   (k) 1090 km;   (l) 1150 m.

4   To how many significant figures have the following measurements been given?
   (a) 3 cm;   (b) 4·1 cm;   (c) 4·10 cm;   (d) 6·00 m;
   (e) 105 km;   (f) 88·5 km;   (g) 7·13 km;   (h) 2·01 cm.

D 5   To how many significant figures are the following figures stated?
   (a) The circumference of the earth is 40 000 000 metres.
   (b) The height of Mount Everest is 8882 m.
   (c) The population of the world in 1960 was 2 717 350 000.

D 6   (a) If you were timing a 100 m sprint to what accuracy would you want to be able to give the time? (To the nearest unit, or tenth, or hundredth...of a second?)

   (b) If you are measuring an angle with a protractor, to what accuracy can you give your answer in degrees?

   (c) To what accuracy can you weigh flour on ordinary kitchen scales?

   (d) 'Her wage is about £16 per week.' Is it likely that the '£16' has been rounded off to the nearest unit?

   (e) 'He is 12 years old and is already 130 cm tall.' Have these quantities been rounded off to the nearest unit?

   (f) A pair of shoes is marked at £2·95. What is this rounded off to the nearest pound? Why are the extra 5p not added to the price?

Rounding off and significance

2   The reason for the ridiculous answer will be obvious to the pupils, but as has been suggested in the text there is no *one* sensible answer.

$$£693·47 = £700 \text{ (to 1 s.f.)}$$
$$= £690 \text{ (to 2 s.f.)}$$
$$= £693 \text{ (to 3 s.f.),}$$
$$7283·14 = 7000 \text{ km (to 1 s.f.)}$$
$$= 7300 \text{ km (to 2 s.f.)}$$
$$= 7280 \text{ km (to 3 s.f.),}$$
$$49·37 \text{ kg} = 50 \text{ kg (to 1 s.f.)}$$
$$= 49 \text{ kg (to 2 s.f.).}$$

3   (*a*) 100, 140;   (*b*) 200, 180;   (*c*) 900, 950;
    (*d*) £60, £63;   (*e*) 10 km, 12 km;   (*f*) 20 m, 23 m;
    (*g*) 20 km, 15 km;   (*h*) £300, £300;
    (*i*) £7000, £7000;   (*j*) 1000 km, 1000 km;
    (*k*) 1000 km, 1100 km;   (*l*) 1000 m, 1200 m.

4   (*a*) 1;   (*b*) 2;   (*c*) 3;   (*d*) 3;
    (*e*) 3;   (*f*) 3;   (*g*) 3;   (*h*) 3.

5   (*a*) In this case, 4 is probably the only significant figure. However, it should be made clear that, out of this context and unless stated to the contrary, some of the zeros in a measure such as 40 000 000 *could be* significant.
    (*b*) 4.
    (*c*) 6, for the population of the world could not be estimated to within one thousand.

6   (*a*) It would depend to a large extent on the importance of the event; at school, readings rounded off to the nearest tenth might be sufficient, but at national or international level, hundredths of a second are required.
    (*b*) To the nearest degree.
    (*c*) To within 30 – 50 g.
    (*d*) Yes.
    (*e*) It is most likely, but we cannot be certain.
    (*f*) £3. £2·95 'sounds' a lot less than £3!

Interlude

NUMBER NAMES

There are two main advantages that number names have over word names. One lies in the fact that numbers are *ordered*. (Letters are ordered in their arrangement in the alphabet but, when two or more letters are required, this order becomes less obvious.) It is the ordering that matters in numbering the houses in a road, rooms in a hotel, platforms on a station, pegs in a row, etc. Having seen two of the numbers, you 'know where you are' within the set.

Numbers that are used in this way can be meaningfully subtracted in order to find out how many houses, rooms or pegs there are between the ones you are looking for and the ones you are looking at. Otherwise operations between these numbers are generally meaningless.

The second advantage in the use of number names is that they provide labels for a great many people or objects and that these labels are sure to be different from one another and this lessens the likelihood of mistaken identity. For example, each man in an army or each person working for a County Authority may be given a number and so be easily distinguished for purposes of transfer, payment, etc., even though they have the same name. This advantage also applies to manufactured articles which need to be distinguished from each other.

When referring to car numbers, pupils might mention insurance, accident identification, proof of possession or the difficulty in identifying a car in a car park. Similarly, for valuable or complicated articles they may mention the need for identifying numbers for insurance, guarantee or replacement purposes.

Such public, but anonymously uniformed, people as policemen or bus conductors carry numbers so that they may be identified by the public who may wish to compliment them on (or complain about) the work they do.

Interlude

Pupils may argue the usefulness of road numbers as opposed to lists of towns in describing a route. They may also have noticed that a road leading from a main route sometimes has an associated number. From London, it is evident that the roads were originally numbered clockwise from the A. 1 to the North to the A. 6. Roads such as the A. 303 or the A. 10 lead from the main A. 3 or A. 1. Much of the original system has been destroyed by the formation of new roads.

Another subject which may be discussed is the advantage or disadvantage of all numeral telephone numbers.

At present, each person whose birth is registered in this country is provided with a number so that they may be identified under the National Health Service.

Interlude

NUMBER NAMES

In this book, all the pages, chapters, sections, figures, questions and examples are numbered. As the numbers are always attached to the same pages, figures, etc., they can be used as *names* for the pages and figures. We refer to 'page 137' or to 'Figure 3'.

Write about things outside the school which you know are given number names. Explain why they are often more useful than word names. Can number names ever be combined, for example, by addition or subtraction?

Here are some suggestions that you can use to get you started:

House numbers. How are house numbers arranged along a street? Is the same arrangement always used? Do the two sides of a street ever get out of order?

Car numbers. Could all the numbers be replaced by letters? Does the order in which numbers are attached to cars matter?

Why are there numbers on some peoples' uniforms? (For example, those of bus conductors and policemen.)

Road numbers. Are there other ways of numbering roads? Are streets in a town numbered? Look at a map of the main roads in England. Can you see any order in the way the roads coming out of big cities are numbered?

Valuable manufactured articles. Very often things like bicycles, cameras, clocks and trains are numbered. What are the uses of these number names?

Have you a number name?

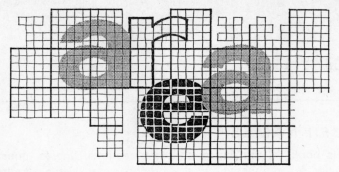

4. Area

1. COMPARISON OF AREA

Can you tell which of these two shapes is the larger?

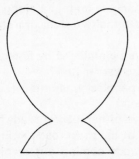 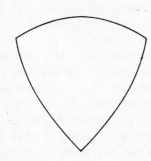

Fig. 1

If the shapes are nearly the same size, it is not easy to tell which is the larger just by looking at them. We need a way of *comparing* the areas. Discuss how you might do this.

You could, for example, see how many small coins of the same size you could fit inside each shape (see Figure 2). This *might* give you a rough guide. How?

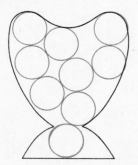 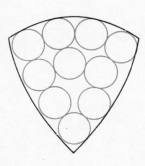

Fig. 2

4. Area

As a result of the work in the Prelude and the chapter on tessellations, pupils will already be familiar with the idea that the plane can be 'filled' with certain shapes and not with others. We now use this work as a basis for the development of the important idea that area is fundamentally a measure of a region of the plane. It is the appreciation of area and the often accompanying difficulties of calculating it that are of primary importance, and not the rules for finding the areas of certain geometrical figures.

In the last section of this chapter, the area of a rectangle is discussed. Pupils are then encouraged to discover and use the relation

$$\text{area} = \text{length} \times \text{breadth},$$

with, we hope, a full understanding of its meaning. In *Book C*, this work will be extended to the discovery of rules for finding the areas of triangles and parallelograms. Again, it will be the derivation and application of the rules that are important, not the remembering of them.

In this course we attempt to give the pupil the understanding that will enable him to quickly find and adapt formulae for himself, should he need them. Certain of the basic formulae are to be found in books of tables and we consider that these should always be available.

1. COMPARISON OF AREA

Comparison of sizes by eye is not easy when the shapes are dissimilar and the areas differ by small amounts.

An attempt to estimate by covering the surface with anything available will help. Fivepenny pieces, pennies, ends of pencils, rubbers, drawing pins, sweets or anything else which is reasonably small and flat will do. Large diagrams on the blackboard could be covered by hands.

Since it is possible to fit 10 coins in the right-hand shape and 9 in the left-hand one, this might indicate that the former has the larger area. This happens to be true, but the important thing is that we cannot be *sure* by counting coins.

The coin in Figure 2 is a particularly bad choice because there are gaps between the coins and large parts have to be omitted as the coins chosen are too large for the space to be covered.

Area

In Figure 3, the coin is again too large, because although the area of the shape on the left is 1 cm² larger than the one on the right, coin counting *might* lead one to suppose that the difference was greater than this.

1.1 Tessellations and units of area

It very soon becomes obvious that a unit of area is needed so that results can be compared.

(*b*) The initial difficulty would be in obtaining a circle and a triangle of the same area, and even if this were possible, the shapes to be measured might be more suited to be filled with circles than triangles or vice versa, which would give biased results. In general it is better to use the same shape, although Figures A and B illustrate a case when this is not so.

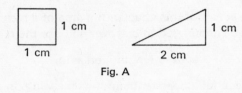

Fig. A

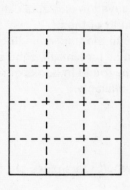
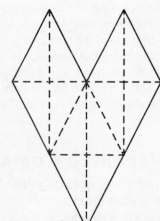

Fig. B

Comparison of area

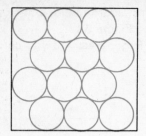 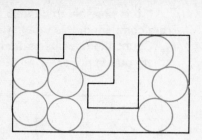

Fig. 3

There are two reasons, however, why the coin shown in Figure 2 was a particularly bad choice for the comparison of these two shapes. What are they?

Look at Figure 3. Count the number of coins fitted inside each shape. Which would you estimate to be the larger? How certain are you?

1.1 Tessellations and units of area

(*a*) In Chapter 3, we found that when comparing the widths of desks, it was necessary for everybody to use the same unit of length. (A 'pencil-length' unit is not much use when the pencils are different sizes.) In a similar way, when you compare the areas of two shapes everyone must use the same unit of area.

(*b*) The units of area must be the same size but they need not be the same shape though this is usually convenient. What would be the difficulty in comparing areas, if one shape was measured with circles and the other with triangles?

(*c*) The unit of area we used in Section 1 was a circle that was 9 mm across. However, we have seen that it was not a good choice. It would have been better to have chosen a shape that left no gaps.

For example, Figure 4 shows the shapes of Figure 1 filled with squares of the same size. At least the squares fit together more tightly.

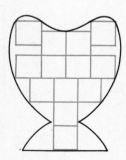 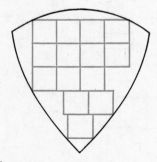

Fig. 4

Area

If it is possible to choose a shape to fit exactly into both the areas we are comparing, this would be better still. (See Figure 5.)

Count the squares to find which has the larger area. Are you certain that you are right?

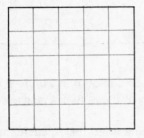 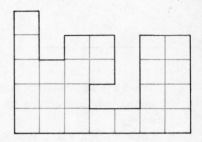

Fig. 5

Exercise A

1 Compare these pairs of shapes. Say which is the larger of each pair.

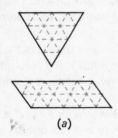 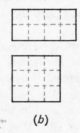 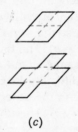

(a)　　　　　　　　(b)　　　　　　　　(c)

2 Count the square tiles to compare the areas of each pair of shapes.

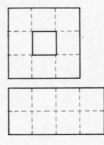
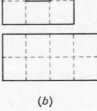
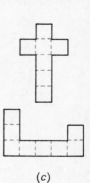

(a)　　　　　　　　(b)　　　　　　　　(c)

Comparison of area

Figure 5 shows the shapes of Figure 3 now filled completely with squares. This time we can be certain that the square has the larger area of the two.

Exercise A

1. (a) 16 units, 20 units; (b) 8 units, 9 units; (c) 4 units, 5 units.

2. (a) 10 units, 9 units; (b) 8 units, 8 units; (c) 7 units, 8 units; (d) 13 units, 15 units.

Area

3 (a) 6 units,   (b) 9 units,
         7 units;   10 units;
   (c) 28 units,   (d) 12 units,
         26 units;   11 units.

Comparison of area

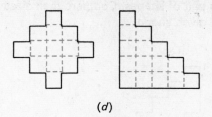

(d)

3 Count the triangular tiles to compare the areas of each pair of shapes.

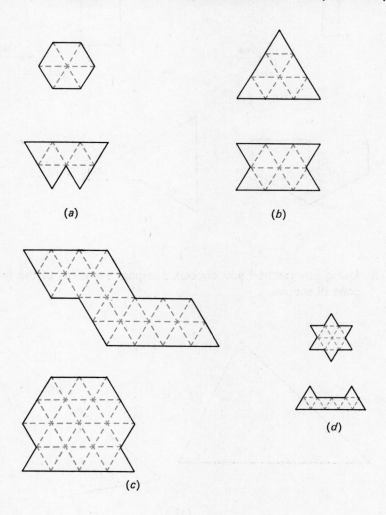

Area

4 Trace each pair of shapes. Compare their sizes by filling them in with tiles of the given shape.

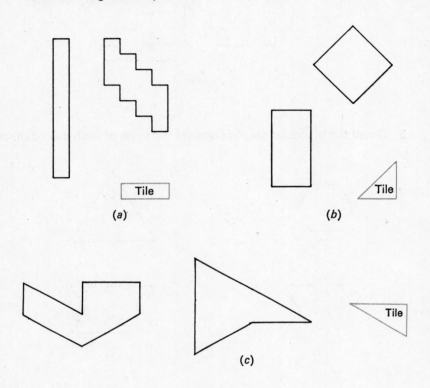

5 Using any method you choose, compare the areas of the following pairs of shapes.

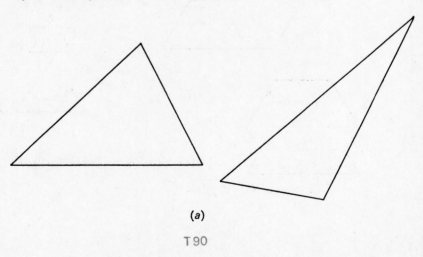

(a)

Comparison of area

4 See Figure C.

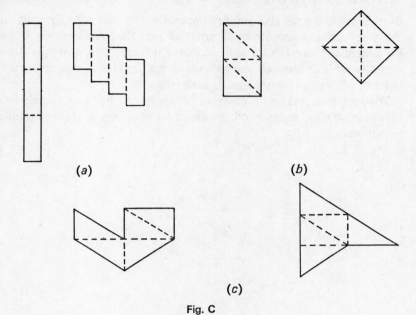

Fig. C

5 (a) The triangle on the left is the larger.
   (b) The cross on the right is the larger, but only fractionally.

Area

2. MEASUREMENT OF AREA

By this stage, pupils will see the necessity, not only for using the same measure in the classroom to compare shapes they have drawn, but also for a standard measure for general use. The actual shape of 'tile' used is a matter for individual choice as long as the final result can be expressed in units which will be universally understood.

The fact that, just as no measure of length can be 'exact', neither can an exact physical measurement of area be obtained, is an important one to put across.

Comparison of area

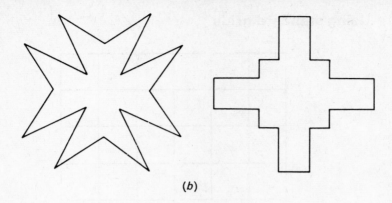

(b)

2. MEASUREMENT OF AREA

When we want to *compare* the areas of two figures, we choose the most convenient unit of area. But we often need to *measure* and describe the area of one figure. To do this, we compare it with some fixed standard unit of area which is familiar to everyone (just as we did for lengths). The *shape* of the standard unit of area can still be chosen for our own convenience. When we state that a certain area is 37 units we give no indication of the shape of the region.

Two common units of area are the square centimetre (cm^2) and the square metre (m^2). The following shapes each have an area of one square centimetre.

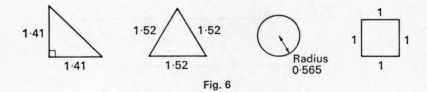

Fig. 6

You can see from the units of length associated with these areas that the square-shaped unit would be the easiest one to draw.

A square metre is roughly the area covered by 24 of these books.

Because it is a process of comparison with a standard unit, any measure of area can only be approximate. For greater accuracy we have to use fractions of the unit area and greater care in the comparison.

Area

2.1 Using standard grids

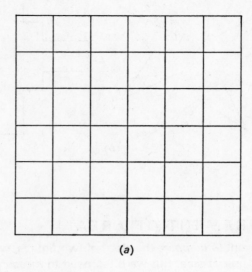

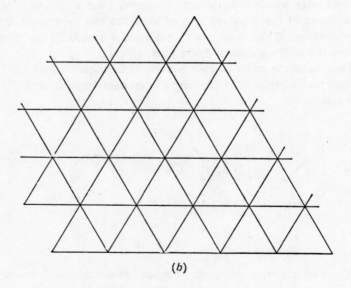

Fig. 7

The squares of Figure 7 (*a*) are one square centimetre in area. So are each of the equilateral triangles of Figure 7 (*b*).

Measurement of area

2.1 Using standard grids

The use of a standard grid simplifies the working considerably and enables the pupils to obtain reasonable estimates of the areas of irregular figures. It is worth recommending that the grid be moved so that the largest possible number of complete squares can be utilized, before counting begins. Estimation of the parts needed to make approximately one unit will need considerable practice with some pupils. However, this should not be insisted on too finely.

Area

Exercise B

1 See Figure D.

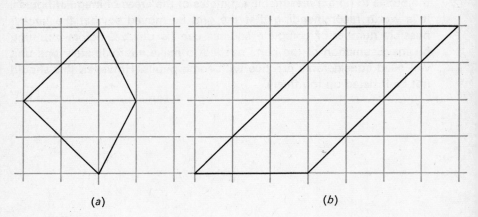

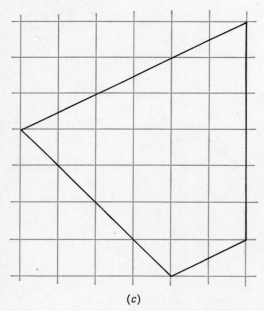

Fig. D

(a) 6 cm²;   (b) 12 cm²;   (c) 24 cm².

Measurement of area

To find the area of an irregular figure, it is convenient to use tracing paper. Either the outline of the figure can be traced and the tracing held over the top of the grid so that the squares can be counted, or the grid can be placed on tracing paper and held over the figure.

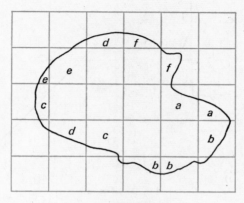

Fig. 8

Figure 8 shows a shape drawn over the square grid. It is possible to count and *estimate* its area from the figure.

How many complete centimetre squares are there?

```
The complete squares have area    7
Two parts marked a                1
Three parts marked b              1
Two parts marked c                1
Two parts marked d                1
Two parts marked e                1
Two parts marked f
  together with the small
  unlettered parts                1
                                 ___
                                 13 squares
```

Exercise B

1 Draw each of the following quadrilaterals on centimetre graph paper:

   (a) (2, 0), (0, 2), (2, 4), (3, 2);
   (b) (0, 1), (3, 1), (7, 5), (4, 5);
   (c) (4, 0), (6, 1), (6, 7), (0, 4).

Find the area of each figure by counting squares.

Area

2

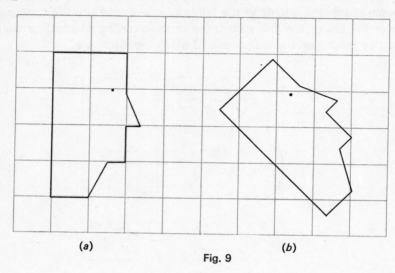

Fig. 9

The area of the polygonal man in Figure 9(a) is approximately $7\frac{1}{2}$ cm². Another polygonal man is shown in Figure 9(b). Has he the same area? Count squares to check your answer.

3 Draw a circle of radius 4 cm on graph paper and estimate the area enclosed.

Draw another circle with radius 2 cm. Would you expect the area of this circle to be half that of the first circle? Check your answer.

4 Use tracing paper and centimetre graph paper to compare the areas of the following pairs of English counties:

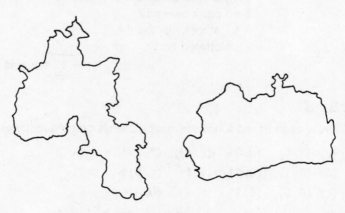

Fig. 10(a)

Measurement of area

2   This question provides an excellent example of the benefit to be obtained by careful positioning of the grid. Figure 9(b) is comparatively difficult to estimate.

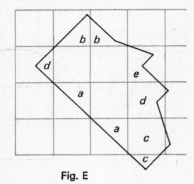

Fig. E

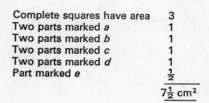

Complete squares have area	3
Two parts marked a	1
Two parts marked b	1
Two parts marked c	1
Two parts marked d	1
Part marked e	$\frac{1}{2}$
	$7\frac{1}{2}$ cm²

Without actually counting squares of side 1 mm this is the best approximation which can be made.

Area

3

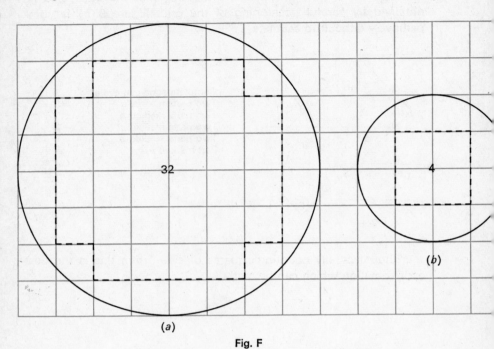

Fig. F

Area of circle radius 4 cm is 50·27 cm².

Area of circle radius 2 cm is 12·57 cm².

It is not expected that the pupils will be able to obtain answers nearer than 50 cm² and 13 cm². An answer of 12 cm² would also be acceptable.

Measurement of area

4 (a) Oxfordshire 1937 km² (748 ml²); Surrey 1867 km² (721 ml²).
   (b) Durham 2626 km² (1014 ml²); Leicestershire 2134 km² (824 ml²).

Area

6   It will quickly be seen that many different areas may be enclosed with any given loop. Theoretically the smallest area is a rectangle of length equal to half the string and of no width. The physical conditions of using string make this impossible but very small areas can be achieved. The largest area is enclosed by forming the loop into a circle. The equilateral triangle is the triangle which encloses the largest area.

3. AREAS OF RECTANGLES

Once the idea of area as a measure of surface has been established, we can progress to the standard work on rectangular shapes.

Measurement of area

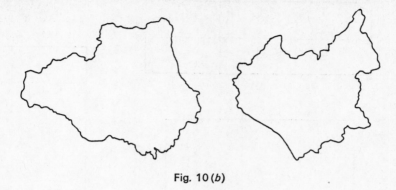

Fig. 10 (b)

Try to identify the counties. You can then check your answers from Whittaker's Almanac.

5   Use centimetre graph paper to discover the areas of irregular shapes of your own choice.

6   Take a length of string and knot it to form a loop. Arrange the loop to form different shapes. Have they all the same area? If not, what is the smallest area you can enclose? What shape encloses the largest area? What shape of triangle is the largest that can be made?

3. AREAS OF RECTANGLES

Look at the rectangle below.

Figure 11 (a) has been completely filled with 12 squares of side 1 cm. We say that its area is 12 cm².

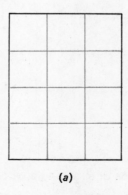

(a)

Fig. 11

T103

Area

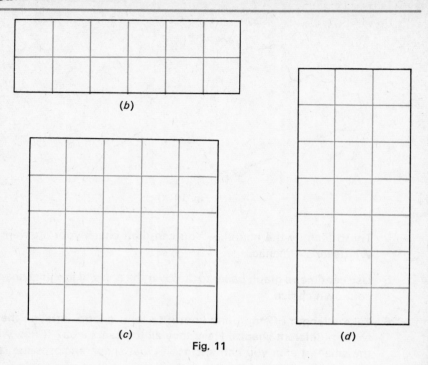

Fig. 11

Figures 11 (*b*), (*c*) and (*d*) have also been filled with centimetre squares. Count the squares to find the areas of these rectangles.

Look again at the dimensions (the width and length) of the rectangle. What is the connection between these and the area in each case?

(*a*) State the area (in cm^2) of the rectangle in Figure 12 without first counting squares?

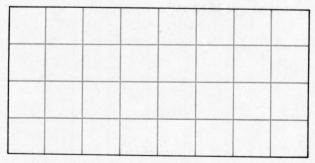

Fig. 12

(*b*) What is the area of a rectangle whose length and width are:

    (i) 11 cm and 4 cm;   (ii) 6 cm and 8 cm?

Areas of rectangles

The areas of the rectangles in Figure 11 are: (*b*) 12 cm²; (*c*) 25 cm²; (*d*) 21 cm².

Most pupils will very quickly see that the product of the dimensions will give them the area, but for those who waver, a reminder of the way rectangles were identified in the work on number pattern (*Book A*, Chapter 1) may be found useful.

(*a*) 32 cm².
(*b*) (i) 44 cm²; (ii) 48 cm².
(*c*) 7 cm, 4 cm; 28 cm².
(*d*) $a \times b$ (or ab) cm².

Most of the examples that pupils have met so far have involved figures drawn to size. The change over from this, to figures drawn to scale might be a possible cause of confusion unless carefully explained.

Area

Exercise C

1. (*a*) (i) 12 cm²; (ii) 252 cm²; (iii) 6 km².
   (*b*) (i) 100 cm²; (ii) 196 cm²; (iii) 900 cm².
   (*c*) (i) 14 cm; (ii) 20 cm; (iii) 12 km.
   (*d*) (i) 5 cm; (ii) 10 m; (iii) 1 km. Don't be tempted to mention square roots!

2. (*c*) has the largest area and (*b*) the smallest.

Areas of rectangles

(c) Measure the sides of the rectangle in Figure 13 to find its area.

Fig. 13

(d) A rectangle has length a cm and width b cm (where a and b are counting numbers). What is its area?

Exercise C

1 (a) What is the area of a rectangle whose sides are:
   (i)   3 cm and 4 cm;
   (ii) 12 cm and 21 cm;
   (iii) 2 km and 3 km?

  (b) What are the areas of squares whose sides are:
   (i)   10 cm;
   (ii) 14 cm;
   (iii) 30 cm?

  (c) What is the length of rectangles where areas and widths are:
   (i)   28 cm² and 2 cm;
   (ii) 340 cm² and 17 cm;
   (iii) 120 km² and 10 km?

  (d) What is the length of a side of a square whose area is:
   (i)   25 cm²;
   (ii) 100 m²;
   (iii) 1 km²?

2 Three rectangles have the following dimensions:
  (a) 9 m and 6 m;
  (b) 3 m and 17 m;
  (c) 4 m and 14 m.

  Which has the largest area and which the smallest?

Area

3 What are the areas of the rectangles whose vertices are given by the following coordinates:

(a) (0, 0), (7, 0), (0, 3), (7, 3);
(b) (1, 2), (5, 2), (1, 6), (5, 6);
(c) (8, 1), (13, 1), (8, 8), (13, 8)?

4 The following figures have been split up into rectangles. Find their areas.

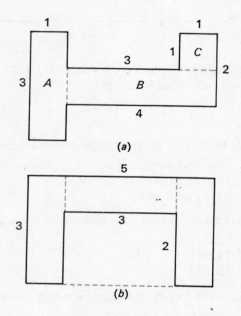

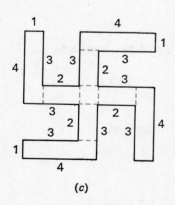

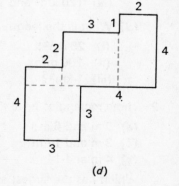

Fig. 14

Areas of rectangles

3 (a) 21;
    (b) 16;
    (c) 35.

4 (a) 8;   (b) 9;   (c) 25;   (d) 28.

Area

5   Pupils should realize that there is more than one way of splitting up these shapes.

   (a) 9;   (b) 40;   (c) 245.

   Questions 6, 7 and 8 are more difficult and need not be attempted by the less able at this stage.

6   (a) 240;   (b) 42.

7   30; 4 m 80 cm (4·8 m).

8   3; No. There are many examples of rectangles which do not fit. The possible dimensions of the large rectangles are:

60×1; 30×2; 20×3; 15×4; 12×5; 10×6.

10×2 and 5×4 will not fit into 60×1.
20×1 and 5×4 will not fit into 30×2.
10×2 and 5×4 will not fit into 20×3.
20×1 and 10×2 will not fit into 15×4.
20×1 and 10×2 will not fit into 12×5.
20×1 and 5×4 will not fit into 10×6.

Areas of rectangles

5 Split the following shapes into rectangles and find the area of each.

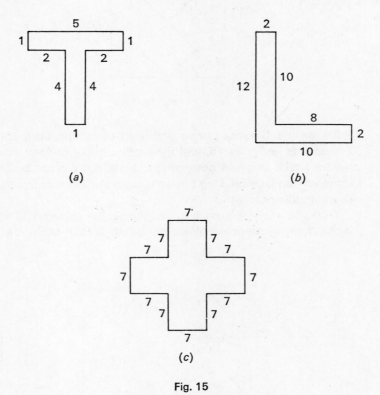

Fig. 15

6 How many tiles with dimensions 5 cm and 12 cm will be needed to fit a space whose dimensions are:
   (a) 100 cm and 144 cm;
   (b) 70 cm and 36 cm?

7 A rectangular floor of length 6 m is being covered by square tiles of side 20 cm. How many tiles will be needed to make a strip one tile wide the length of the room? If 720 tiles are needed altogether, what is the width of the room?

8 How many rectangles of area 20 cm² can be fitted into a rectangle of area 60 cm²? Can this be done for any shape of rectangle? If not, give examples of rectangles of area 20 cm² which do not fit.

T111

Area

9

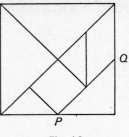

Fig. 16

An ancient Chinese puzzle consisted of cutting up a square into 7 pieces as in Figure 16, and then either trying to form them into a square again, or some other shape. Mark out a 5 cm square in the same way and, by pricking through, make the 7 pieces in card. (*P* and *Q* are middle points.)

Try to form the pieces into shapes similar to those in Figure 17 without any overlapping. What will be the area of each new shape?

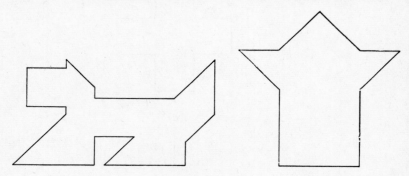

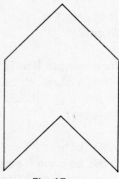

Fig. 17

Areas of rectangles

9 See Figure G. The formation of designs known as Tangrams, by means of these seven pieces was a favourite pastime of the Chinese, who produced many volumes of designs representing all sorts of familiar objects. Figure H shows some further designs for those pupils who are interested. (They might well be asked to form the rectangle.) Others might like to record designs of their own.

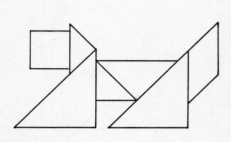
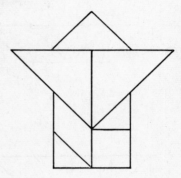
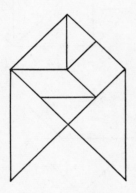

Fig. G

Area

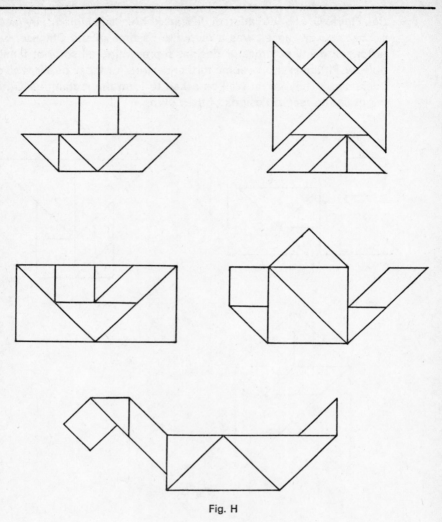

Fig. H

5. Comparison of fractions

This chapter attempts to extend the pupil's understanding of fractions by concentrating on equivalence of fractions—the idea central to almost any manipulation of these numbers, but an idea which is often insufficiently stressed.

The aim is to give pupils an understanding of the meaning of equivalence by visual methods—there is a great deal of drawing and plotting of points. This plotting can be done either on centimetre square paper or on spotty paper. The techniques of adding and subtracting fractions are not dwelt upon at any length—these follow automatically given a good grounding in the basic idea of equivalence.

Comparison of fractions

1. REPRESENTING FRACTIONS

(*a*) It is interesting to invite suggestions from the class as to different ways of representing fractions. A little time spent on discussing the merits of each, is well worth while. In fact most pupils seem to prefer the method of representation used in this book, that is to use two lines, the *T* line across the page and the *B* line up the page. Many books (including the original *S.M.P. Book I*) place the axes the other way round.

The argument may be stated in this way. If the denominator is represented across the page, the fraction gives the gradient of the graph according to the accepted convention. On the other hand, it is also a custom to place the first mentioned number of a pair across the page; this is true of coordinates and vectors and, we feel, should also be true of fractions.

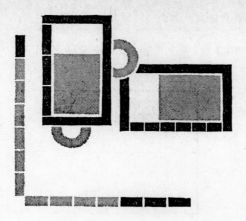

5. Comparison of fractions

1. REPRESENTING FRACTIONS

(a) In Book A, we showed the counting numbers as points on a line, as in Figure 1.

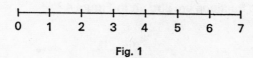

Fig. 1

Fractions are made up of two counting numbers? How can they be represented? How would *you* show $\frac{1}{4}$, $\frac{1}{2}$, $\frac{3}{4}$ and other fractions?

Perhaps you might decide to show them on two number lines, with the top number on one and the bottom number on the other. They could be joined with a line, like this

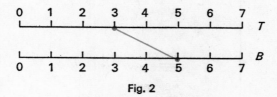

Fig. 2

What fraction do you think is shown in Figure 2?

Can you think of a better position for the *T* and *B* number lines than having them one above the other?

Comparison of fractions

(b) Try drawing the *B* line at right-angles to the *T* line, meeting at 0, as in Figure 3.

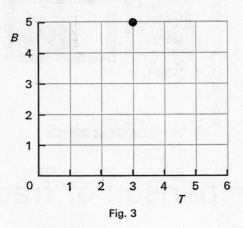

Fig. 3

Figure 3 shows $\frac{3}{5}$ again, but in a much neater way than is shown in Figure 2. Where do you remember seeing a different kind of number pair plotted in this way? We shall refer to this way of representing fractions as *graphing* the fractions.

(c) What fractions are shown in Figure 4?

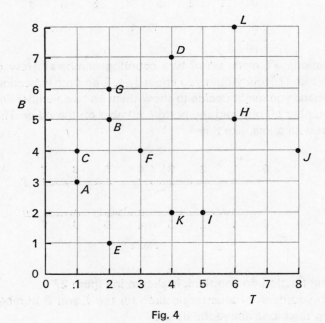

Fig. 4

Representing fractions

(b) We can plot both $\frac{3}{5}$ and (3, 5) in the same way as they are both *ordered pairs*.

(c) $A(\frac{1}{3})$, $B(\frac{2}{5})$, $C(\frac{1}{4})$, $D(\frac{4}{7})$, $E(\frac{2}{7})$, $F(\frac{3}{4})$,
$G(\frac{2}{6})$, $H(\frac{6}{5})$, $I(\frac{5}{2})$, $J(\frac{8}{4})$, $K(\frac{4}{2})$, $L(\frac{6}{8})$.

Comparison of fractions

2. EQUIVALENT FRACTIONS

(*a*) In each case, the points representing the fractions in the set lie on a straight line through the origin.

The fractions in a set each represent the same part of a unit. They are equivalent.

(*b*) A coin could be used for drawing the circles.

Each picture shows the same amount of shaded area.

The plotted points all lie on a straight line through the origin.

If more practice in the idea of equivalence is required, then the following are suggested:

(i) The fraction wall.

Fig. A

Draw a series of strips as in Figure A, on a sheet of plain paper. Make each strip 12 cm long. Divide the first strip into two parts of equal length so that each part represents a half; divide the second strip into three equal parts so that each part represents a third and continue to do this at least until you have strips divided to represent tenths.

Representing fractions

(d) Draw the two number lines *T* and *B* as shown in Figure 4 and graph the following fractions:

(i) $\frac{2}{3}$, (ii) $\frac{5}{7}$, (iii) $\frac{4}{3}$, (iv) $\frac{5}{6}$, (v) $\frac{4}{5}$, (vi) $\frac{3}{2}$.

2. EQUIVALENT FRACTIONS

(a) Draw the lines *T* and *B* and graph the set of fractions:

$\{\frac{3}{4}, \frac{6}{8}, \frac{9}{12}, \frac{12}{16}\}$.

What do you notice?
Now graph the set

$\{\frac{1}{3}, \frac{2}{6}, \frac{3}{9}, \frac{4}{12}\}$.

What do you notice now?
Look carefully at the fractions in each set. What can you say about them?

(b) Draw five small circles of the same size. Shade $\frac{1}{2}$ of the first circle, $\frac{2}{4}$ of the second, $\frac{3}{6}$ of the third, $\frac{4}{8}$ of the fourth, and $\frac{5}{10}$ of the fifth circle. What do you notice about the shaded sections?

The fractions: $\frac{1}{2}, \frac{2}{4}, \frac{3}{6}, \frac{4}{8}, \frac{5}{10}$ are called 'equivalent' fractions.

Graph this set of equivalent fractions. What do you notice about the points? Did you expect this?

Exercise A

1 What fraction of each figure is shaded?

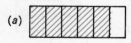 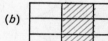

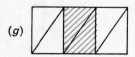 

Comparison of fractions

2. State some sets of equivalent fractions from the answers to Question 1.

3. What fraction of each of the following figures is shaded? Give your answers in two ways as two equivalent fractions.

For example 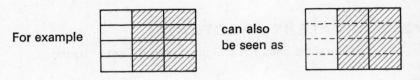 can also be seen as

so $\frac{8}{12}$ and $\frac{2}{3}$ are equivalent fractions.

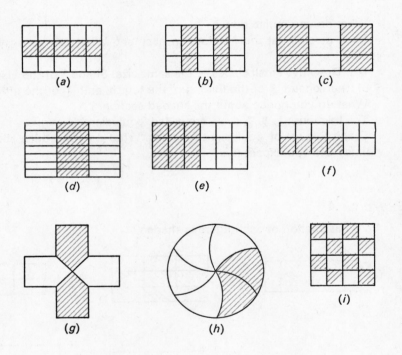

4. Trace each of the following figures, and by drawing more lines yourself, give two equivalent fractions for each of the shaded sections.

For example, by drawing one more line 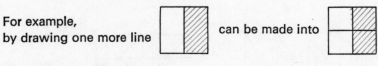 can be made into

so $\frac{1}{2}$ and $\frac{2}{4}$ are equivalent.

T122

Equivalent fractions

Discuss such questions as:

(a) What methods can be used for dividing a section into (i) fifths; and (ii) sevenths?

(b) Do the 'outer' marks get closer and closer to the outside edges? Do these dividing marks form a pattern which has line symmetry? Do you think that the bottom points of the 'outer' marks could be made to lie on a smooth curve?

(c) Do any of the marks lie directly above those of another strip? What does this signify? Make out lists of equivalent fractions that are illustrated by the figure.

(ii) Angles and time on clock faces.

It may be necessary to start with some questions such as the following:

(a) How many minutes are there in an hour? What fraction of an hour is a minute?

(b) Through what fraction of a whole turn does the minute hand move in:

   (i) 1 minute,
   (ii) 5 minutes,
   (iii) 15 minutes,
   (iv) 40 minutes?

(c) In how many minutes does the minute hand move through 45°, 60°, 270°, or 300°?

Write down the equivalent fractions mentioned in answering these questions. For example: a quarter of an hour shows the equivalence of $\frac{15}{60}$ (minutes); $\frac{1}{4}$ (fractions of a whole turn); $\frac{90}{360}$ (degrees).

Exercise A

1. (a) $\frac{5}{6}$;   (b) $\frac{3}{9}$ (or $\frac{1}{3}$);   (c) $\frac{2}{4}$ (or $\frac{1}{2}$);
   (d) $\frac{3}{6}$ (or $\frac{1}{2}$);   (e) $\frac{6}{18}$ (or $\frac{1}{3}$);   (f) $\frac{6}{12}$ (or $\frac{1}{2}$);
   (g) $\frac{2}{6}$ (or $\frac{1}{3}$);   (h) $\frac{4}{8}$ (or $\frac{1}{2}$).

2. (i) $\{\frac{2}{4}, \frac{3}{6}, \frac{4}{8}, \frac{6}{12}\}$;   (ii) $\{\frac{1}{3}, \frac{2}{6}, \frac{3}{9}, \frac{6}{18}\}$.

3. (a) $\frac{5}{15}$ or $\frac{1}{3}$;   (b) $\frac{6}{15}$ or $\frac{2}{5}$;   (c) $\frac{6}{9}$ or $\frac{2}{3}$;
   (d) $\frac{7}{21}$ or $\frac{1}{3}$;   (e) $\frac{9}{18}$ or $\frac{1}{2}$;   (f) $\frac{4}{12}$ or $\frac{1}{3}$;
   (g) $\frac{2}{4}$ or $\frac{1}{2}$;   (h) $\frac{2}{6}$ or $\frac{1}{3}$;   (i) $\frac{8}{16}$ or $\frac{1}{2}$.

Comparison of fractions

4 (a) $\frac{1}{4}$ equivalent to $\frac{2}{8}$, $\frac{3}{12}$, etc.;   (b) $\frac{4}{7}$ equivalent to $\frac{8}{14}$, $\frac{12}{21}$, etc.;
   (c) $\frac{2}{5}$ equivalent to $\frac{4}{10}$, $\frac{6}{15}$, etc.;   (d) $\frac{2}{3}$ equivalent to $\frac{4}{6}$, $\frac{6}{9}$, etc.;
   (e) $\frac{1}{2}$ equivalent to $\frac{2}{4}$, $\frac{3}{6}$, etc.;   (f) $\frac{3}{4}$ equivalent to $\frac{6}{8}$, $\frac{9}{12}$, etc.

5 Both sets of points lie on straight lines through the origin.

6 (a) $\frac{8}{12}$;   (b) $\frac{4}{14}$;   (c) $\frac{1}{4}$;   (d) $\frac{4}{5}$.

7 This is most easily done by placing a ruler on the page, pivoting it at the origin, and turning it until two or more points appear on the same straight line.

   The sets of equivalent fractions are: $\{\frac{1}{3}, \frac{2}{6}\}$, $\{\frac{3}{4}, \frac{6}{8}\}$, $\{\frac{2}{1}, \frac{4}{2}, \frac{8}{4}\}$.

8 (a) No;   (b) no;   (c) yes;
   (d) no;   (e) yes;   (f) no;
   (g) yes;   (h) yes.

Equivalent fractions

(a)   (b)  (c)

(d)  (e)  (f)

5 Graph the sets of equivalent fractions you found in Question 2. What do you notice about each set of plotted points?

6 In each of the following sets there are three equivalent fractions. Graph the fractions, see which one does not lie on the line through the origin, and write down the odd man out.

(a) $\{\frac{1}{3}, \frac{2}{6}, \frac{3}{9}, \frac{8}{12}\}$; (b) $\{\frac{1}{4}, \frac{2}{8}, \frac{3}{12}, \frac{4}{14}\}$;
(c) $\{\frac{5}{10}, \frac{4}{8}, \frac{3}{6}, \frac{1}{4}\}$; (d) $\{\frac{2}{3}, \frac{4}{5}, \frac{6}{9}, \frac{8}{12}\}$.

7 Find the sets of equivalent fractions in Figure 4.

8

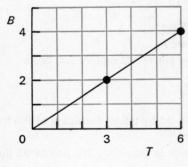 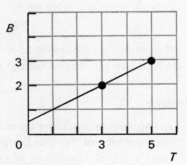

Fig. 5

The line passes through 0, so $\frac{3}{2}$ and $\frac{6}{4}$ are equivalent.   The line does not pass through 0, so $\frac{3}{2}$ and $\frac{5}{3}$ are *not* equivalent.

By graphing the following pairs of fractions as in Figure 5, see if they are equivalent to each other:

(a) $\frac{1}{2}, \frac{3}{4}$; (b) $\frac{3}{4}, \frac{9}{16}$; (c) $\frac{4}{7}, \frac{8}{14}$;
(d) $\frac{3}{4}, \frac{2}{3}$; (e) $\frac{3}{12}, \frac{1}{4}$; (f) $\frac{7}{21}, \frac{1}{4}$;
(g) $\frac{3}{8}, \frac{6}{16}$; (h) $\frac{4}{8}, \frac{1}{2}$.

Comparison of fractions

2.1 Forming equivalent fractions

Graph the set of fractions

$$\{\tfrac{2}{3},\ \tfrac{4}{6},\ \tfrac{6}{9},\ \tfrac{8}{12},\ \tfrac{10}{15}\}.$$

Are the members of this set equivalent to each other?

How is the set of numbers on the tops of the fractions made up? Can you do the same for the set of numbers on the bottoms of the fractions? Can you write down some more members of this set of equivalent fractions? Graph the new fractions to see if they are correct.

In fact the tops of the fractions are multiples of two and the bottoms are multiples of three, like this

$$\left\{\frac{1\times 2}{1\times 3},\ \frac{2\times 2}{2\times 3},\ \frac{3\times 2}{3\times 3},\ \frac{4\times 2}{4\times 3},\ \frac{5\times 2}{5\times 3}\right\}.$$

So the next fraction is $\dfrac{6\times 2}{6\times 3}$, that is, $\dfrac{12}{18}$.

Write down some more members of the set

$$\{\tfrac{3}{4},\ \tfrac{6}{8},\ \tfrac{9}{12},\ \tfrac{12}{16},\ \ldots\}.$$

What multiples are being used for the top and bottom numbers of the fractions? Graph the fractions you have written down to see whether they are in fact equivalent.

Write down some more members of the set

$$\{\tfrac{2}{5},\ \tfrac{4}{10},\ \ldots\}.$$

What multiples are being used here?

Write down some members of the set of equivalent fractions

$$\{\tfrac{3}{7},\ \ldots\}.$$

Summary

Sets of equivalent fractions are made up by using multiples of the top and bottom numbers of the first fraction.

When a set of equivalent fractions is graphed, the points all lie on a straight line through the origin.

Exercise B

Write down a set of at least three fractions which are equivalent to each of the following fractions. Write them in two ways each time, so that, for example, $\tfrac{2}{5}$ gives

$$\left\{\frac{1\times 2}{1\times 5},\ \frac{2\times 2}{2\times 5},\ \frac{3\times 2}{3\times 5},\ \frac{4\times 2}{4\times 5}\ \ldots\right\}$$

or

$$\{\tfrac{2}{5},\ \tfrac{4}{10},\ \tfrac{6}{15},\ \tfrac{8}{20},\ \ldots\}.$$

T126

Equivalent fractions

2.1 Forming equivalent fractions
Yes.

The answer expected is that the top is the '2 times table' and the bottom the '3 times table'. The phrase 'multiple of 2' is used in the chapter rather than the familiar '2 times table' but the two phrases can be used together until the meaning of the new one is firmly understood.

The set $\{\frac{3}{4}, \frac{6}{8}, \frac{9}{12}, \frac{12}{16}, \frac{15}{20}\}$ uses the multiples of 3 and 4,

the set $\{\frac{2}{5}, \frac{4}{10}, \frac{6}{15}, \frac{8}{20}\}$ the multiples of 2 and 5, and

the set $\{\frac{3}{7}, \frac{6}{14}, \frac{9}{21}\}$ the multiples of 3 and 7.

Exercise B

1   (a) $\{\frac{3}{7}, \frac{6}{14}, \frac{9}{21}, \frac{12}{28}, ...\};$ (b) $\{\frac{4}{5}, \frac{8}{10}, \frac{12}{15}, \frac{16}{20}, ...\};$

    (c) $\{\frac{3}{2}, \frac{6}{4}, \frac{9}{6}, \frac{12}{8}, ...\};$ (d) $\{\frac{5}{3}, \frac{10}{6}, \frac{15}{9}, \frac{20}{12}, ...\}.$

2   (a) $\{\frac{1}{2}, \frac{2}{4}, \frac{3}{6}, \frac{4}{8}, ...\};$ (b) $\{\frac{5}{6}, \frac{10}{12}, \frac{15}{18}, \frac{20}{24}, ...\};$

    (c) $\{\frac{4}{7}, \frac{8}{14}, \frac{12}{21}, \frac{16}{28}, ...\};$ (d) $\{\frac{7}{9}, \frac{14}{18}, \frac{21}{27}, \frac{28}{36}...\}.$

3   (a) $\{\frac{4}{3}, \frac{8}{6}, \frac{12}{9}, \frac{16}{12}, ...\};$ (b) $\{\frac{5}{7}, \frac{10}{14}, \frac{15}{21}, \frac{20}{28}, ...\}.$

Comparison of fractions

3. COMPARING FRACTIONS

A picture such as Figure B is what is wanted.

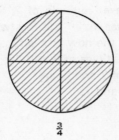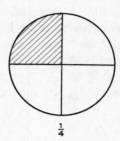

$\frac{3}{4}$ $\frac{1}{4}$

Fig. B

Fractions which have the same denominator are ordered by their numerators. To compare (or order) fractions with different denominators, we must first find new fractions, equivalent to each, but with the same denominator.

We feel that this process should be understood before arithmetic processes involving 'lowest common multiples' are mentioned.

Exercise C

1. (a) $\{\frac{4}{5}, ..., \frac{12}{15}\}$, $\{\frac{2}{3}, ..., \frac{10}{15}\}$, $\frac{4}{5} > \frac{2}{3}$;
   (b) $\{\frac{1}{2}, ..., \frac{7}{14}\}$, $\{\frac{4}{7}, \frac{8}{14}\}$, $\frac{4}{7} > \frac{1}{2}$;
   (c) $\{\frac{3}{4}, ..., \frac{15}{20}\}$, $\{\frac{4}{5}, ..., \frac{16}{20}\}$, $\frac{4}{5} > \frac{3}{4}$;
   (d) $\{\frac{7}{8}, ..., \frac{21}{24}\}$, $\{\frac{5}{6}, ..., \frac{20}{24}\}$, $\frac{7}{8} > \frac{5}{6}$.

2. The line of fractions equivalent to the bigger one is always nearer to T line.

3. (a) $\{\frac{1}{3}, ..., \frac{5}{15}\}$, $\{\frac{2}{5}, ..., \frac{6}{15}\}$, $\frac{2}{5} > \frac{1}{3}$;
   (b) $\{\frac{1}{4}, \frac{2}{8}\}$, $\{\frac{3}{8}, ...\}$, $\frac{3}{8} > \frac{1}{4}$;
   (c) $\{\frac{4}{5}, ..., \frac{28}{35}\}$, $\{\frac{5}{7}, ..., \frac{25}{35}\}$, $\frac{4}{5} > \frac{5}{7}$;
   (d) $\{\frac{5}{8}, ..., \frac{15}{24}\}$, $\{\frac{7}{12}, \frac{14}{24}\}$, $\frac{5}{8} > \frac{7}{12}$.

3.1 Calculating equivalent fractions

It is hoped that by now pupils should be able to take short cuts in working out equivalent fractions and this exercise is to help establish these short cuts.

Equivalent fractions

Plot each set to see if you are correct.

1. (a) $\frac{3}{7}$; (b) $\frac{4}{5}$; (c) $\frac{3}{2}$; (d) $\frac{5}{3}$.
2. (a) $\frac{1}{2}$; (b) $\frac{5}{6}$; (c) $\frac{4}{7}$; (d) $\frac{7}{9}$.
3. (a) $\frac{4}{3}$; (b) $\frac{5}{7}$.

3. COMPARING FRACTIONS

Which is bigger, $\frac{3}{4}$ or $\frac{1}{4}$? Draw a picture to show your answer.
Which is bigger, $\frac{2}{3}$ or $\frac{1}{3}$? Again draw a picture.
Which is bigger, $\frac{2}{5}$ or $\frac{3}{5}$? Do you need a picture in order to decide?
All these questions are easy to answer. But can you tell which is bigger, when the fractions are $\frac{5}{6}$ or $\frac{3}{4}$? Why is this question more difficult to answer than the previous ones?
The sets of fractions equivalent to $\frac{5}{6}$ and $\frac{3}{4}$ are written out below:

$$\{\tfrac{5}{6}, (\tfrac{10}{12}), \tfrac{15}{18}, \tfrac{20}{24}, \ldots\},$$
$$\{\tfrac{3}{4}, \tfrac{6}{8}, (\tfrac{9}{12}), \tfrac{12}{16}, \ldots\}.$$

It is clear that $\frac{10}{12} > \frac{9}{12}$, so it follows that $\frac{5}{6} > \frac{3}{4}$.

Exercise C

1. Write out the sets of fractions equivalent to each member of the following pairs. Pick out the two fractions with the same bottom number in order to see which fraction is bigger.

   (a) $\frac{4}{5}, \frac{2}{3}$; (b) $\frac{1}{2}, \frac{4}{7}$; (c) $\frac{3}{4}, \frac{4}{5}$; (d) $\frac{7}{8}, \frac{5}{6}$.

2. Graph your answers to Question 1 and see if the line of fractions equivalent to the bigger one is always nearer one of the axes. If it is, then this is another check on your working.

3. Write out the sets of fractions equivalent to each member of the following pairs, and plot sets of points to check that they are correct. Pick out the two fractions with the same bottom number in order to see which is bigger.
   Check to see if the line of fractions equivalent to the bigger one is closer to the *T* number line.

   (a) $\frac{1}{3}, \frac{2}{5}$; (b) $\frac{1}{4}, \frac{3}{8}$; (c) $\frac{4}{5}, \frac{5}{7}$; (d) $\frac{5}{8}, \frac{7}{12}$.

3.1 Calculating equivalent fractions

In Exercise C you were looking for the smallest number that appeared among the multiples of both the bottom numbers of the fractions you

Comparison of fractions

were comparing. If you could go through the multiples tables in your head, or even guess the smallest number appearing in both multiples tables, then of course this would be much quicker.

Exercise D

1 Copy and fill in the missing parts:

(a) $\dfrac{2}{3} = \dfrac{4 \times 2}{4 \times 3} = \dfrac{}{12}$;

(b) $\dfrac{3}{4} = \dfrac{5 \times 3}{5 \times 4} = \dfrac{}{20}$;

(c) $\dfrac{2}{5} = \dfrac{3 \times 2}{3 \times 5} = \dfrac{}{15}$;

(d) $\dfrac{3}{5} = \dfrac{}{4 \times 5} = \dfrac{}{20}$;

(e) $\dfrac{4}{7} = \dfrac{}{3 \times 7} = \dfrac{}{21}$;

(f) $\dfrac{5}{6} = \dfrac{}{} = \dfrac{}{24}$;

(g) $\dfrac{4}{9} = \dfrac{}{} = \dfrac{}{54}$;

(h) $\dfrac{3}{10} = \dfrac{}{} = \dfrac{}{60}$;

(i) $\dfrac{7}{11} = \dfrac{}{} = \dfrac{}{33}$;

(j) $\dfrac{5}{12} = \dfrac{}{} = \dfrac{}{60}$.

3.2 Graphing fractions to compare them

Example 1

Compare $\frac{2}{3}$ and $\frac{3}{4}$.

The smallest number appearing in both the multiples of three and multiples of four is 12; so

$$\tfrac{2}{3} = \tfrac{8}{12} \quad \text{and} \quad \tfrac{3}{4} = \tfrac{9}{12}.$$

Since $\tfrac{9}{12} > \tfrac{8}{12}$, it follows that $\tfrac{3}{4} > \tfrac{2}{3}$.

Check:

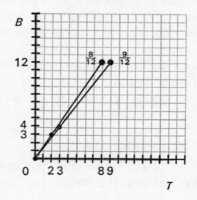

Comparing fractions

Exercise D

1. (a) $\dfrac{2}{3} = \dfrac{4\times 2}{4\times 3} = \dfrac{8}{12};$ (b) $\dfrac{3}{4} = \dfrac{5\times 3}{5\times 4} = \dfrac{15}{20};$

   (c) $\dfrac{2}{5} = \dfrac{3\times 2}{3\times 5} = \dfrac{6}{15};$ (d) $\dfrac{3}{5} = \dfrac{4\times 3}{4\times 5} = \dfrac{12}{20};$

   (e) $\dfrac{4}{7} = \dfrac{3\times 4}{3\times 7} = \dfrac{12}{21};$ (f) $\dfrac{5}{6} = \dfrac{4\times 5}{4\times 6} = \dfrac{20}{24};$

   (g) $\dfrac{4}{9} = \dfrac{6\times 4}{6\times 9} = \dfrac{24}{54};$ (h) $\dfrac{3}{10} = \dfrac{6\times 3}{6\times 10} = \dfrac{18}{60};$

   (i) $\dfrac{7}{11} = \dfrac{3\times 7}{3\times 11} = \dfrac{21}{33};$ (j) $\dfrac{5}{12} = \dfrac{5\times 5}{5\times 12} = \dfrac{25}{60}.$

Comparison of fractions

3.2 Graphing fractions to compare them
Exercise E

1. (a) $\frac{3}{8} > \frac{1}{3}$; (b) $\frac{1}{4} < \frac{2}{5}$;
   (c) $\frac{2}{7} < \frac{3}{10}$; (d) $\frac{4}{6} > \frac{7}{9}$;
   (e) $\frac{5}{12} > \frac{3}{10}$.

2. (a) $\frac{7}{6}$ or $1\frac{1}{6}$; (b) $\frac{11}{12}$;
   (c) $\frac{1}{6}$; (d) $\frac{5}{12}$;
   (e) $\frac{11}{10}$ or $1\frac{1}{10}$; (f) $\frac{23}{18}$ or $1\frac{5}{18}$.

3. (a) $3\frac{1}{6}$; (b) $\frac{7}{12}$;
   (c) $\frac{3}{6}$ or $\frac{1}{2}$; (d) $\frac{37}{18}$ or $2\frac{1}{18}$;
   (e) $\frac{67}{15}$ or $4\frac{7}{15}$; (f) $\frac{35}{24}$ or $1\frac{11}{24}$.

3.3 The fraction number line

This section is intended to provoke class discussion on number systems. The following points may be raised.

(1) In theory, the more integral points that are extended over the plane, the more lines there will be to cut R and so the more fraction points there will be along that line. Their number can be increased without limit. In practice, of course, there are limitations caused by thickness of the lines drawn.

(2) Although there are an infinite number of fractions (rational numbers) they do not completely cover the number line.

The set of all positive fractions (that is fractions equal to or greater than zero) represents the set of numbers called 'positive rational numbers'. Within this set, every equation of the form

$$ax = b \quad (a \neq 0 \text{ and } a, b \text{ rational})$$

has a solution (cf. *Teacher's Guide for Book A,* Chapter 9).

Comparing fractions

Example 2

Fractions with different bottom numbers can be added and subtracted using these ideas. Find the sum of $\frac{2}{3}$ and $\frac{1}{5}$.

The smallest number appearing in both the multiples of three and the multiples of five is 15.

So
$$\frac{2}{3} = \frac{10}{15} \text{ and } \frac{1}{5} = \frac{3}{15}.$$
$$\frac{2}{3} + \frac{1}{5} = \frac{10}{15} + \frac{3}{15} = \frac{13}{15}.$$

Exercise E

1. Using the method of Example 1 compare:

    (a) $\frac{3}{8}$ and $\frac{1}{3}$; (b) $\frac{1}{4}$ and $\frac{2}{5}$;
    (c) $\frac{2}{7}$ and $\frac{3}{10}$; (d) $\frac{4}{6}$ and $\frac{7}{9}$;
    (e) $\frac{5}{12}$ and $\frac{3}{10}$.

2. Using the method of Example 2, work out:

    (a) $\frac{1}{2} + \frac{2}{3}$; (b) $\frac{3}{4} + \frac{1}{6}$;
    (c) $\frac{5}{6} - \frac{2}{3}$; (d) $\frac{2}{3} - \frac{1}{4}$;
    (e) $\frac{1}{2} + \frac{3}{5}$; (f) $\frac{5}{6} + \frac{4}{9}$.

3. When dealing with mixed numbers, all that is necessary is to turn them into fractions. For example:

$$1\frac{1}{3} + 2\frac{1}{2},$$
$$1\frac{1}{3} = \frac{4}{3} = \frac{8}{6} \text{ and } 2\frac{1}{2} = \frac{5}{2} = \frac{15}{6},$$

so
$$1\frac{1}{3} + 2\frac{1}{2} = \frac{8}{6} + \frac{15}{6} = \frac{23}{6} = 3\frac{5}{6}.$$

Now do the following:

   (a) $1\frac{1}{2} + 1\frac{2}{3}$; (b) $2\frac{1}{4} - 1\frac{2}{3}$;
   (c) $1\frac{1}{6} - \frac{2}{3}$; (d) $1\frac{5}{6} + \frac{2}{9}$;
   (e) $2\frac{2}{3} + 1\frac{4}{6}$; (f) $2\frac{5}{8} - 1\frac{1}{6}$.

3.3 The fraction number line

In Exercise C, you noticed that the larger the fraction the closer its line of equivalent fractions was to the *T* line. Let us look into this a little more closely.

Comparison of fractions

In Figure 6 the sets of equivalent fractions have been drawn for $\frac{1}{2}, \frac{1}{1}, \frac{2}{1}, \frac{3}{1}, \frac{4}{1}, \frac{5}{1}$. Where do these lines meet the line R?

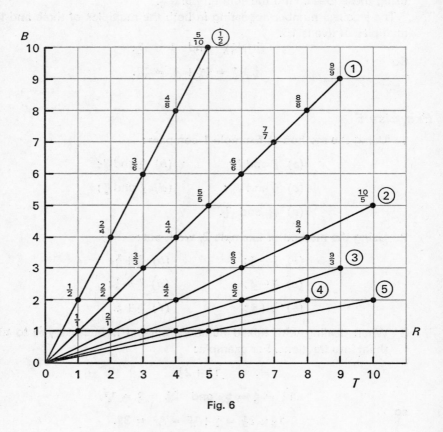

Fig. 6

In Figure 7, we have enlarged part of Figure 6 to show some more of the equivalent fraction lines that can be drawn. Look carefully at the places where they meet the line R. Why do you think that the line R can be thought of as the number line for fractions?

Make as large a drawing as you can and see how many fractions you can show on the number line R. Why can't you show any more? Are there any more fractions to graph?

How many fractions do you think there are on the number line? Does the complete set of fractions entirely cover the number line, or are there any gaps left?

Comparing fractions

When represented on the number line, the rational numbers are spread out 'infinitely finely': every interval, however small, contains infinitely many rational numbers. But, even so, the rationals do not completely fill the number line. There are several ways of showing this:

(i) If we take for granted that rational numbers can be represented by decimals which either terminate or recur, then any non-terminating non-recurring decimal will represent a number which is not rational. (Examples are 0·110100100010... and 0·123571113171923....)

(ii) Consider the number $\sqrt{2}$. This cannot be expressed in the form m/n, for, if it were, we should have $2n^2 = m^2$ and, assuming that m and n are prime to each other and that any numbers have a unique set of prime factors, this equation cannot be true, for the left-hand side has an odd number of prime factors, whereas the right-hand side has an even number.

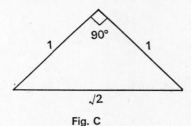

Fig. C

Numbers like $\sqrt{2}$ can evidently (see Figure C) be represented by a length of line; so can non-terminating non-recurring decimals. They can be represented on the number line and yet are not rational numbers so the rational numbers do not completely fill the number line.

(3) The numbers that fill in the gaps in the number line are called irrational numbers.

Comparison of fractions

The fractions graphed in Figure 8 are $\frac{1}{3}, \frac{1}{4}, \frac{2}{3}, \frac{3}{2}, \frac{4}{4}$. In placing the ruler on the grid and letting it pass through the origin, we are marking the lines of equivalent fractions. As the ruler rotates towards the *T* line, it uncovers the fractions in order of size, since the larger the fraction, the nearer to the *T* line is the line of its equivalent fractions.

Exercise F

1 $\frac{1}{4}, \frac{1}{3}, \frac{2}{5}, \frac{4}{7}, \frac{2}{3}, \frac{3}{4}, \frac{6}{5}, \frac{2}{1}, \frac{5}{2}$.

   A and G are equivalent, and so are E, K and J.

2 (a) $\frac{1}{6}, \frac{2}{9}, \frac{1}{4}$; (b) $\frac{4}{7}, \frac{3}{5}, \frac{2}{3}$;
   (c) $\frac{2}{3}, \frac{3}{4}, \frac{5}{6}, \frac{6}{7}$; (d) $\frac{3}{7}, \frac{4}{9}, \frac{5}{11}, \frac{5}{8}, \frac{7}{10}$.

Comparing fractions

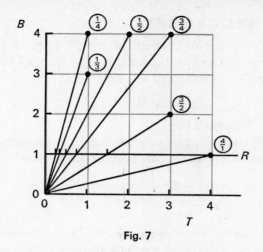

Fig. 7

What fractions are graphed in Figure 8? Now take a ruler, place it along the *B* line and turn it slowly about (0, 0) in a clockwise direction. As the ruler turns, it will uncover fractions in order of size, smallest first. Why?

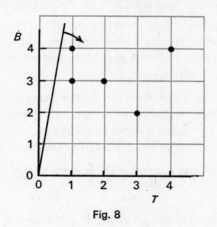

Fig. 8

Exercise F

1 Look back at Figure 4 and with the help of a ruler, write down all the fractions shown there, in order of size, smallest first. What do you notice about the fractions *A* and *G*? Why does this happen? Are there any more like this in the figure?

2 Arrange the following sets of fractions in order of size:

(a) $\frac{1}{4}$, $\frac{1}{6}$, $\frac{2}{9}$;

(b) $\frac{3}{5}$, $\frac{4}{7}$, $\frac{2}{3}$;

(c) $\frac{2}{3}$, $\frac{6}{7}$, $\frac{5}{6}$, $\frac{3}{4}$;

(d) $\frac{5}{11}$, $\frac{4}{9}$, $\frac{3}{7}$, $\frac{5}{8}$, $\frac{7}{10}$.

Comparison of fractions

4. FAREY SEQUENCES

The Farey sequence of fractions of order 4 is the set of all simplified fractions less than one, whose bottom numbers are 4 or less, arranged in order of size. The members of this set are

$$\tfrac{1}{2}, \tfrac{1}{3}, \tfrac{2}{3}, \tfrac{1}{4}, \tfrac{3}{4}.$$

What has happened to $\tfrac{2}{4}$?

Graph the fractions, and with the help of a ruler arrange them in size, smallest first. Now find the Farey sequences of order 5 and 6 in the same way.

Exercise G

1 Write down the Farey sequence of order 7. Consider any three consecutive members. Find out how to combine the two outer ones to form the middle one. Check that the same rule applies to all triples (that is, sets of 3 consecutive numbers).

2 Consider any pair of consecutive members of the Farey sequence of order 4. Multiply the top number of the second by the bottom number of the first. Multiply the top number of the first by the bottom number of the second. Subtract your answers. Repeat this for another pair. What do you find? Check to see whether this is true in Farey sequences of order 5, 6 and 7.

*3 Find the number of members in the Farey sequences of orders 3, 4 and 5. Is there a general rule? Estimate the number for the sequence of order 6 and check it. Now check the number for the Farey sequence of order 7 which you obtained in Question 2. Explain what you find.

*4 Three consecutive members of a Farey sequence are

$$\ldots, \tfrac{1}{3}, \tfrac{3}{8}, \tfrac{2}{5}, \ldots$$

Find the next member, bearing in mind your answers to Questions 1 and 2.

4. FAREY SEQUENCES

Farey sequences have applications only in advanced mathematics but they are included because many pupils enjoy dealing with them. They are, moreover, valuable in giving some more practice in graphing and ordering fractions.

Exercise G (The 'starred' questions are more difficult.)

1 $\frac{1}{7}, \frac{1}{6}, \frac{1}{5}, \frac{1}{4}, \frac{2}{7}, \frac{1}{3}, \frac{2}{5}, \frac{3}{7}, \frac{1}{2}, \frac{4}{7}, \frac{3}{5}, \frac{2}{3}, \frac{5}{7}, \frac{3}{4}, \frac{4}{5}, \frac{5}{6}, \frac{6}{7}.$

Taking any three consecutive elements, say $\frac{1}{3}, \frac{2}{5}, \frac{3}{7}$, adding the tops of the outer pairs and the bottoms of the outer pairs gives the middle element, thus
$$\frac{1+3}{3+7} = \frac{4}{10} = \frac{2}{5}.$$

2 Taking any two consecutive elements from a Farey sequence, say $\frac{5}{7}, \frac{3}{4}$, we find
$$5 \times 4 = 20,$$
$$7 \times 3 = 21, \text{ difference } 1.$$

The result is generally true.
The answers to Questions 1 and 2 are not independent.
Consider the sequence
$$\frac{a}{b}, \frac{c}{d}, \frac{e}{f};$$
$$bc - ad = 1 \quad \text{and} \quad de - cf = 1,$$
$$bc - ad = de - cf,$$
$$bc + cf = de + ad,$$
$$c(b + f) = d(a + e),$$
$$\frac{c}{d} = \frac{a + e}{b + f}.$$

Thus $2 \Rightarrow 1$ (but not $1 \Rightarrow 2$).

Comparison of fractions

3* The numbers of members in the sequences of orders, 3, 4 and 5 are respectively 3, 5 and 9. For each increase in the order, the number of members is increased by the number of proper fractions with the new order as denominator. Thus the number of members for order 6 is $9+2 = 11$, since $\frac{1}{6}$ and $\frac{5}{6}$ are the only proper fractions with denominator 6 (the members already present in the sequence of order 5, namely $\frac{1}{3}$, $\frac{1}{2}$ and $\frac{2}{3}$, eliminate $\frac{2}{6}$, $\frac{3}{6}$ and $\frac{4}{6}$) ; and the number of members for order 7 is $11+6 = 17$ since $\frac{1}{7}$, $\frac{2}{7}$, $\frac{3}{7}$, $\frac{4}{7}$, $\frac{5}{7}$ and $\frac{6}{7}$ are all proper fractions.

Farey sequences

4* If the next member is x/y, the conditions of Question 1 give the equation
$$\frac{3+x}{8+y} = \frac{2}{5}$$
and this simplifies to $5x - 2y = 1$, the condition of Question 2. From these equations the next term might be any of the following:

$$\tfrac{1}{2}, \tfrac{3}{7}, \tfrac{5}{12}, \tfrac{7}{17}, \tfrac{9}{22}, \tfrac{11}{27}, \ldots$$

Each term of this sequence is less than the term on its left. We would obviously choose as small a term as possible and the choice must depend upon the order of the Farey sequence that we have been given. We know that it must be at least of order 8 and so $\tfrac{3}{7}$ is the first choice. But could the sequence be of higher order? This we can discover by trying sequences of higher orders starting with 9 to discover the first sequence which has an element lying between the given terms. $\tfrac{4}{11}$ lies between $\tfrac{1}{3}$ and $\tfrac{2}{5}$ and so the sequence must be of order less than 11 and the term $\tfrac{5}{12}$ would not be acceptable. The answer is, therefore, $\tfrac{3}{7}$.

Revision exercises

Quick quiz, no. 1

1. $\frac{8}{24}$.
2. 54 m.
3. (a) 41·1; (b) 50·1.
4. 22 cm².
5. (a), (b) and (d) tessellate. Regular pentagons and decagons will not tessellate, but some non-regular ones will. (See Prelude, Experiment 5.)
6. $\frac{1}{4}, \frac{1}{2}, \frac{3}{5}, \frac{2}{3}, \frac{5}{6}$.

Quick quiz, no. 2

1. £4·92.
2. 340.
3. The square.
4. 4 cm².
5. {2, 3}, {2, 5}, {2, 7}, {3, 5}, {3, 7}, {5, 7}.
6. (a) True; (b) false; (c) true.

Revision exercises

Quick quiz, no. 1

1. Complete the fraction $\frac{?}{24}$ which is equivalent to $\frac{1}{3}$.

2. What is the perimeter of a regular hexagon of side 9 m?

3. Calculate:
   (a) $\begin{array}{r} 23\cdot 4 \\ +\ 17\cdot 7 \\ \hline \end{array}$
   (b) $\begin{array}{r} 149\cdot 2 \\ -\ \ 99\cdot 1 \\ \hline \end{array}$

4. What is the area of a rectangle of length $5\frac{1}{2}$ cm and width 4 cm?

5. Which of the following figures tessellate:
   (a) regular hexagon;
   (b) right-angled triangle;
   (c) pentagon;
   (d) quadrilateral;
   (e) decagon?

6. Arrange in order of size, smallest first:

   $\frac{2}{3}, \frac{3}{5}, \frac{1}{4}, \frac{5}{6}, \frac{1}{2}$.

Quick quiz, no. 2

1. A book costs £1·23. How much would 4 books cost?

2. Round off 337·52 to 2 significant figures.

3. If A = {quadrilaterals} and B = {regular polygons}, what is $A \cap B$?

4. What is the area of a square whose perimeter is 8 cm?

5. List the subsets of {2, 3, 5, 7} which have just two members.

6. Are the following true or false?
   (a) $\frac{5}{8} > \frac{4}{7}$;
   (b) $\frac{5}{6} < \frac{9}{11}$;
   (c) $\frac{2}{3} + \frac{3}{4} = \frac{17}{12}$.

T143

Revision exercises

Exercise A

1. State whether or not you would need to count accurately if you were:

    (a) a bank clerk counting out £5 notes to a customer;
    (b) a newspaper reporter wanting to give the size of the crowd at a football match;
    (c) a scorer counting up the runs in an innings;
    (d) a builder deciding how many bricks would be needed to build a block of flats;
    (e) a driver reversing out of a garage who was told 'The road will be clear after the third car'.

2. Calculate the angles marked p, q and r (see Figure 1).

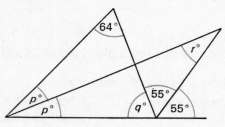

Fig. 1

3. A certain crossword composer always makes his diagrams with rotational symmetry of order 4. He has finished blacking in the squares of the top left-hand quarter of the diagram. Copy and complete Figure 2. How many 'across' clues will he have to invent? How many 'down'?

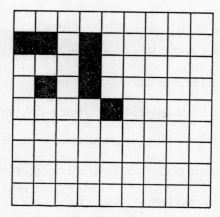

Fig. 2

Revision exercises

Exercise A

1. (a) Yes; (b) no; (c) yes; (d) no; (e) yes.
2. $p = 23, q = 70, r = 32$. (Find q first.)
3.

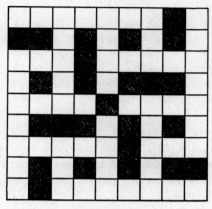

Fig. A

He will have to invent 8 'across' clues and 8 'down' clues.

Revision exercises

4
	(a) Line	Length		(d) Line	Length
	3	2+3		1	2×1
	4	2+4		2	2×2
				3	2×3
(b)	2	2÷2		4	2×4
	3	2÷3		n	2×n
	4	2÷4			
	n	2÷n	(e)	1	2+3
				2	2+4
(c)	3	2+5		3	2+5
	4	2+7		4	2+6
	n	2+(2n−1)		n	2+(n+2)

5 Yes.

(a) 1 (Figure B (a));

(b) 3 (Figure B (b)). (If the triangle is right-angled, the arrowhead degenerates into another right-angled isosceles triangle.) Note that the parallelograms drawn in Figure B (b) are basically the same, as they are mirror images.

(c) 6 (Figure B (c)). (If the triangle is right-angled, the two arrowheads degenerate into right-angled isosceles triangles.)

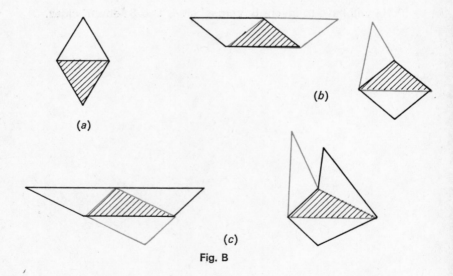

Fig. B

Notice that turning the triangle over only gives more possible quadrilaterals when the triangle is scalene.

It is possible to tessellate *any* quadrilateral because the sum of its angles is 360° (see p. T6).

Revision exercises

4 Complete the following tables. They state:
   (i) the length of each given line segment;
   (ii) an expression for any length of the sequence in terms of *n*. (*n* is any member of the set of counting numbers.)

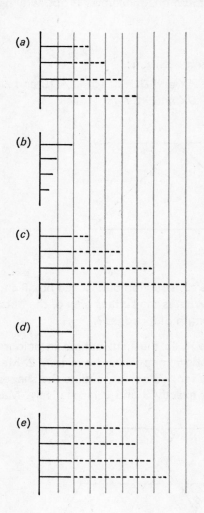

	Line	Length
(a)	1	2+1
	2	2+2
	3	2+
	4	2+
	n	2+n
(b)	1	2÷1
	2	2÷
	3	2÷3
	4	2÷
	n	2÷
(c)	1	2+1
	2	2+3
	3	2+
	4	2+
	n	2+
(d)	1	2
	2	2×2
	3	2
	4	2
	n	2
(e)	1	2
	2	2
	3	2+5
	4	2
	2	n

5 Is it possible to make a tessellation using any shaped triangle as a basic motif?

How many different shaped quadrilaterals can you make by joining edge to edge:

   (a) 2 identical equilateral triangles;

Revision exercises

(b) 2 identical isosceles triangles;

(c) 2 identical scalene triangles (scalene triangles are triangles which have no two sides of the same length)?

Make sketches of your results. Would it be impossible to make a tessellation from any one of these quadrilaterals? If not, why not?

Exercise B

1 In Figure 3, what is the relation between:

(a) AX and AD;   (b) AX and DC;   (c) ∠ADC and ∠DXC;
(d) ∠XDC and DXC;   (e) DX and XC?

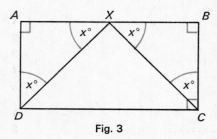

Fig. 3

2 Draw accurately a net for the construction of a closed box 6 cm long, $4\frac{1}{2}$ cm wide and 3 cm high. What is the surface area of the outside of the box? What is the total length of its edges?

3 Make three copies of Figure 4. In the first, outline in colour those shapes which have rotational symmetry of order exactly 2. Mark the centres of rotation. On the other two copies, outline the shapes with rotational symmetry of order exactly 3 and 6 respectively. Mark the centres of rotation.

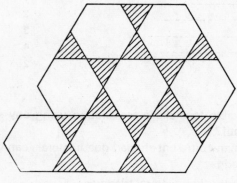

Fig. 4

Revision exercises

Exercise B

1. (a) $AX = AD$; (b) $AX = \frac{1}{2}DC$ or $DC = 2AX$;
   (c) $\angle ADC = \angle DXC$;
   (d) $\angle XDC = \frac{1}{2}\angle DXC$ or $\angle DXC = 2\angle XDC$;
   (e) $DX = XC$.

2. Figure C (b) shows one suitable net for the box.

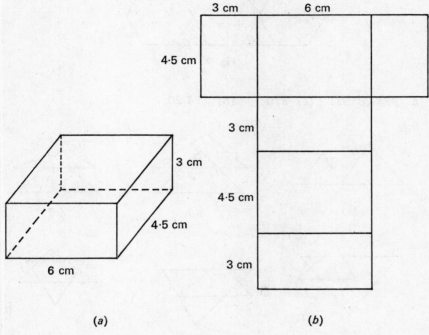

Fig. C

The surface area is $2[(6 \times 4\cdot 5) + (3 \times 4\cdot 5) + (3 \times 6)]$ cm² $= 117$ cm².
The total length of its edges is $4(6 + 4\cdot 5 + 3)$ cm $= 54$ cm.

Revision exercises

3 See Figure D.

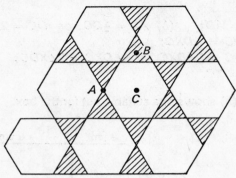

Fig. D

4 374·2015. (a) 370; (b) 374·20.

5

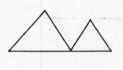

(a)

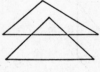

(b)

(c)

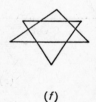

(d)   (e)   (f)

Fig. E

The sides of two triangles can intersect at a maximum of 6 points, the sides of two convex quadrilaterals at a maximum of 8 points and the sides of two convex hexagons at a maximum of 12 points.

Revision exercises

4 Add together 3·87, 0·2485, 0·083 and 370.
   Now write down your answer
   (a) correct to 2 significant figures, and
   (b) correct to 2 decimal places.

5 Make a drawing of two triangles whose sides intersect at
   (a) one point;   (b) two points;   (c) three points;
   (d) four points;   (e) five points;
   (f) six points;   (g) no points.

   What is the greatest number of points at which the sides of 2 triangles can intersect? What is the greatest number of points at which the sides of two quadrilaterals can intersect?

   What is the greatest number of points at which the sides of two hexagons can intersect?

   What is the greatest number of points at which the sides of two n-sided figures can intersect?

6. Angle

1. FIXING A POSITION

A playground conversation (John and Malcolm are talking).

J. 'There you are, that's the boy who won the cross-country last year.'
M. 'Where?'
J. 'Over there, he's the one with the dark hair.'
M. 'I can see lots of boys with dark hair.'
J. 'Well you see the gate, don't you?'
M. 'Yes.'

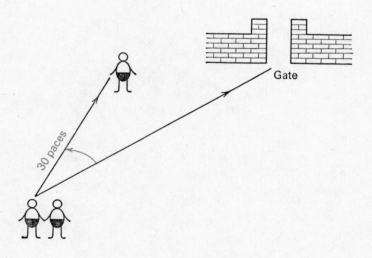

6. Angle

There are two very important points to note in this chapter:
(i) Bearings are always measured *clockwise* from the *north*. (In theoretical mathematical work, particularly in graphs, angles are measured *anticlockwise* from the positive $y = 0$ axis.)
(ii) Bearings are measured on a *horizontal plane*.

1. FIXING A POSITION

Discussion here should include consideration of the locus of possible positions of the boy if only the distance from a fixed point, *or* only the angle turned through from a fixed line, were given.

Angle

2. THE CLOCK-RAY METHOD
Range card

This is a device employed by the Army, and it can be adapted for school use. (Soldiers in a defensive position would make a range-card giving the distance and bearing of places and objects that are likely to become targets.) A school range-card can be made using a roof space or balcony as a vantage point and distances obtained from a local map. Both the Prominent Object and the North Direction method can be used. An example is given in Figure A.

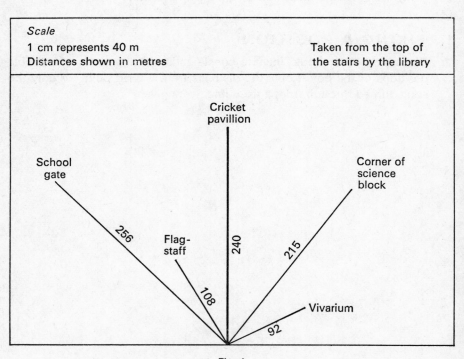

Fig. A

Fixing a position

J. 'Well now turn a little to your left, and you will see that he is about 30 paces away from us.'

M. 'The boy looking down now?'

J. 'Yes, that's him.'

This method of fixing a position depends on naming a landmark first (the gate), then talking of a turn and a distance. It works quite well over a short distance but is not very accurate. Why not?

2. THE CLOCK-RAY METHOD

A useful way of fixing a position, which has been developed in the Army, uses a clockface to indicate the required direction.

Imagine that you are standing at the centre of a clock which can be turned round so that 12 o'clock is in the direction of the landmark. You can then refer to the amount of turn needed to face the object you are talking about by the hours on the clockface. For example:

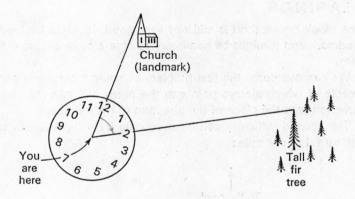

1 Turn the clock round so that 12 faces the landmark.

2 Give the hour which points to the object you are concerned with. In this case you would say to your friend:

'From the Church Steeple, 2 o'clock, Tall Fir Tree.'

Exercise A

Trace the drawing at the top of the next page into your exercise book. Take the top of the church steeple as 12 o'clock. Draw a clockface and give instructions to find all the other objects.

Angle

If one centimetre on this paper represents one hundred metres on the actual ground, say how far away from each object you are.

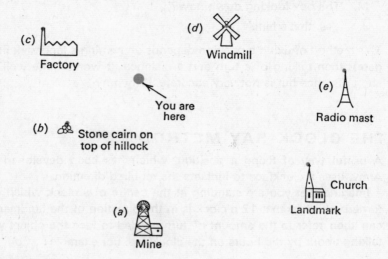

3. BEARINGS

The clock-ray method is still not very good. It might be hard to find a landmark and it might be necessary to give a more accurate measure of turn.

We can overcome the first problem by using a magnetic compass, the needle of which always points to the north. We can then let the north direction take the place of the direction of the landmark.

The second difficulty can be overcome by using degrees to measure the turn. For example:

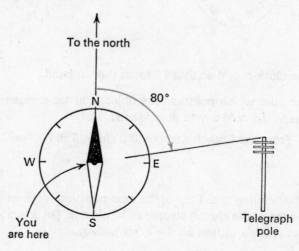

The clock-ray method

Exercise A

A lot of classroom time can be saved and greater accuracy obtained if figures are duplicated and issued to the class to complete. This applies particularly to this exercise and Exercise B, Question 3.

The Church is 510 metres away at 12 o'clock.
- (a) 350 m, 2 o'clock;
- (b) 230 m, 4 o'clock;
- (c) 290 m, 6 o'clock;
- (d) 270 m, 10 o'clock;
- (e) 550 m, 11 o'clock.

3. BEARINGS

Some pupils may be interested to discuss why bearings are measured from, particularly, the 'North' line. The line (or axis) about which the earth rotates cuts the surface of the Earth at what are called the 'North and South Poles'. The Pole Star is very nearly on this line and so appears to be fixed in the heavens while the other stars seem to rotate daily about the axis. The Pole Star gives a fixed direction from which navigators used originally to set their courses when they were in the Northern hemisphere. (If this direction is followed, the North Pole would be reached.) Later they used magnetic compasses (which point in approximately the same direction) and, more recently, the gyro compass to give this fixed direction. It is natural that the North–South line should be used as the 'base' direction when taking bearings as well as setting courses.

It is worth discussing with the class the desirability of putting 080° instead of simply 80°.

It is hoped that mathematics rooms will be built with the compass inlaid in the floor either in metal or in tiles. Compasses painted on the floor are only successful if there is not a great deal of traffic over the area. To overcome this difficulty, the points of the compass can be marked on the ceiling, but this does give rise to some confusion about E. and W.

Angle

Exercise B

1 *A:* 066°, *B:* 164°, *C:* 254°, *D:* 309°.

2 (*a*) 045°;   (*b*) 135°;   (*c*) 315°.

Bearings

In this case you would say to your friend, 'Bearing 080 degrees, telegraph pole.'

BEARINGS:
ALWAYS START FROM NORTH;
ALWAYS TURN CLOCKWISE

Exercise B

1. Use your protractor to measure the bearings of objects at points A, B, C and D, if you are standing at O.

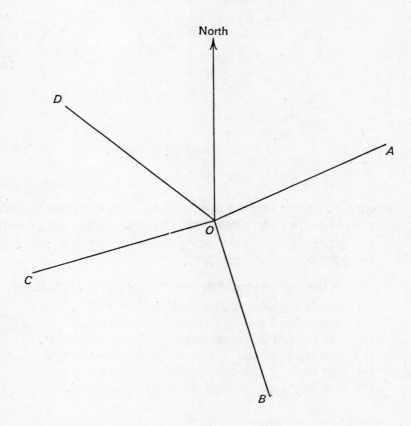

2. What is the bearing of:

(a) north-east;   (b) south-east;   (c) north-west?

Angle

3 Trace the next drawing into your exercise book. Draw a straight line between your position and the red dot on each object, and give its bearing.

Take the scale to be one centimetre to one hundred metres, and say how far away from each object you are.

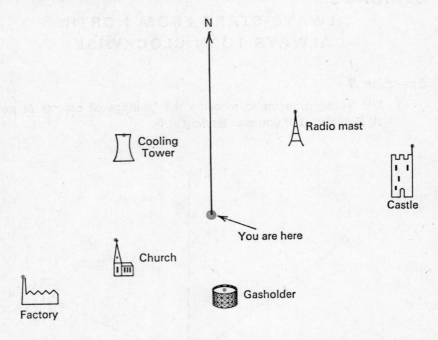

4 Start at a point near the middle of the page and draw a north line straight up the page. Draw lines to indicate the following bearings:

(a) 025°;   (b) 143°;   (c) 235°;
(d) 270°;   (e) 305°.

5 What is the final bearing after obeying each of the following?

(a) Start facing NE and do a clockwise turn of 50°.
(b) Start facing E and do a clockwise turn of 65°.
(c) Start facing SW and do an anticlockwise turn of 45°.
(d) Start facing W and do an anticlockwise turn of 40°.
(e) Start facing S and do a clockwise turn of 100°.
(f) Start facing SE and do a clockwise turn of 140°.

3.1 Plotting a course

Think of a ship about to leave this harbour and sail to the river mouth opposite.

3	Radio mast	360 m, 038°.	Castle	570 m, 072°.
	Gasholder	210 m, 172°.	Church	260 m, 255°.
	Factory	530 m, 251°.	Cooling Tower	320 m, 314°.

5   (a) 095°;   (b) 155°;   (c) 180°;
    (d) 230°;   (e) 280°;   (f) 275°.

3.1 Plotting a course

Scale drawing

It is usually very helpful if a rough sketch is made before any attempt at an accurate drawing. Working from a base-line and a suitable choice of scale is also most important, and various possibilities should be discussed. The use of graph paper makes it easier to find a suitable scale and permits us to draw north lines easily at each change of course.

Angle

In Exercise C, Questions 4, 5 and 6, the idea of back bearings is discussed. The possibility of answering such questions without the use of a protractor should be discussed. If the pupils have any knowledge of parallel lines and alternate angles, the calculation is relatively straightforward, but this knowledge is not part of the course and should not be expected.

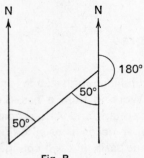

Fig. B

If the bearing of A from B is 050° then the bearing of B from A is 180° + 50° = 230°.

Bearings

The pilot would first look at the compass to find out where north was, and then note the bearing needed to clear the harbour mouth, in this case 070°.

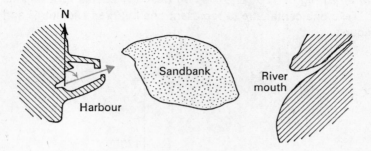

The ship would then sail on a bearing of 070° until it was outside the harbour. To miss the sandbank, the pilot would again find north and change to a bearing of 135°.

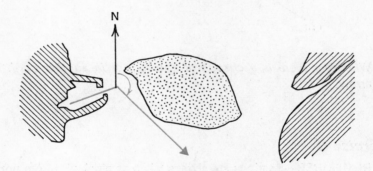

When clear of the sandbank, north would be found once again and a final bearing of 063° would be taken to the river mouth.
  Note that this is only one of many possible courses.

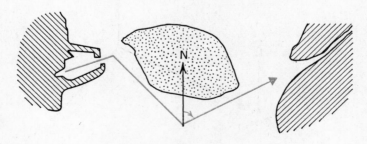

Angle

Example 1

An aircraft flies on a bearing of 055° for 500 km, and then changes course to a bearing of 120° and flies on this new course for 350 km.

Take one centimetre to represent one hundred kilometres and plot this course.

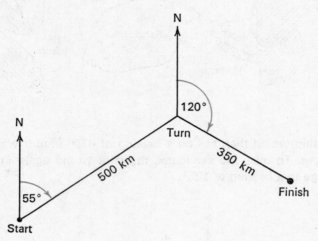

Whenever there is a change of course, put in another north direction arrow.

Exercise C

REMEMBER—Bearings are always taken as clockwise from north.

1. Describe each of these courses. Take the scale to be one centimetre to one hundred kilometres, and north straight up the page.

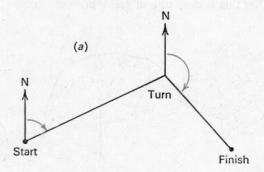

(a)

Bearings

Exercise C
1 (a) 410 km, 063°, followed by 270 km, 140°;

Angle

(b) 440 km, 320°, followed by 390 km, 252°;
(c) 460 km, 229°, followed by 505 km, 087°;
(d) 350 km, 343°, followed by 465 km, 122°;
(e) 400 km, 037°, followed by 420 km, 207°;
(f) 270 km, 090°, followed by 530 km, 270°;

Bearings

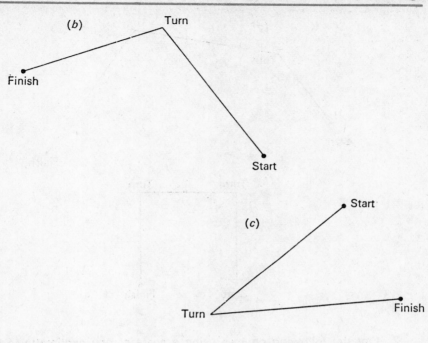

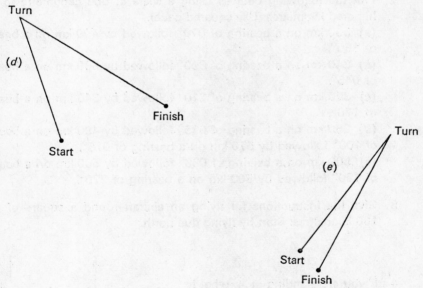

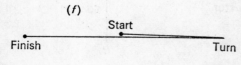

Angle

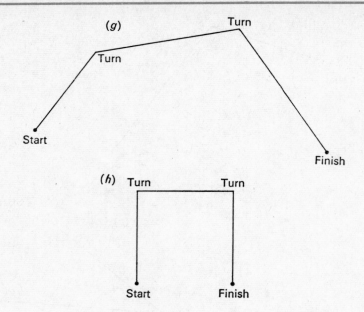

2   Plot the following courses using a scale of one centimetre to one hundred kilometres. Use squared paper.
   (a) 550 km on a bearing of 070° followed by 400 km on a bearing of 150°.
   (b) 340 km on a bearing of 300° followed by 600 km on a bearing of 045°.
   (c) 320 km on a bearing of 210° followed by 340 km on a bearing of 170°.
   (d) 350 km on a bearing of 035° followed by 490 km on a bearing of 100° followed by 570 km on a bearing of 075°.
   (e) 500 km on a bearing of 030° followed by 500 km on a bearing of 150° followed by 500 km on a bearing of 270°.

3   Give the instructions for flying an aircraft round a square of side 100 kilometres. Start by flying due north.

4   If you are standing at A, what is the bearing of B? Can you answer this without using a protractor?

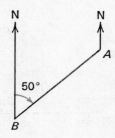

Bearings

(g) 270 km, 043°, followed by 390 km, 084°, followed by 410 km, 150°;

(h) 250 km, 360°, followed by 250 km, 090°, followed by 250 km, 180°.

3 100 km, 360°, followed by 100 km, 090°, followed by 100 km, 180°, followed by 100 km, 270° (in a clockwise direction).

4 230°.

Angle

5 280°.

6 300°.

7 18 km, 125° (see Figure C). Suggested scale for pupils: 1 cm to 5 km or 2 km.

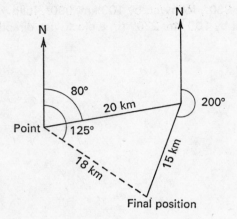

Fig. C

Bearings

8   82 km, 088° (see Figure D).
    Suggested scale for pupils: 1 cm to 10 km or 5 km.

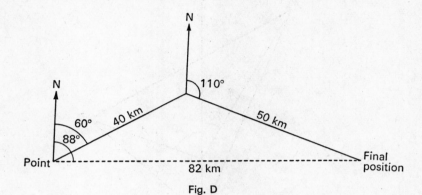

Fig. D

9

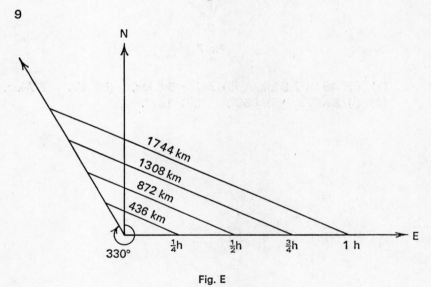

Fig. E

Suggested scale: 1 cm to 100 km.

Angle

10

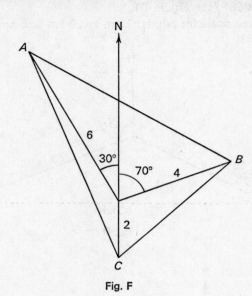

Fig. F

(a) (i) $AB = 7.8$ km, (ii) $BC = 5.5$ km, (iii) $CA = 7.7$ km.
(b) (i) $299°$, (ii) $050°$, (iii) $157°$.

Bearings

5   What is the bearing of P from Q?

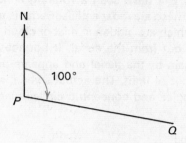

6   The bearing of a town A from a town B is 120°. What is the bearing of town B from town A?

7   A ship leaves port and sails for 20 km on a bearing of 080°; it then alters course to a bearing of 200° and sails for a further 15 km. Make a scale drawing of the complete track, and state the distance and bearing of the ship from the port.

8   A ship leaves port on a bearing of 060°. After 40 km it alters course to a bearing of 110° which it keeps for 50 km. By means of a scale drawing find the new position of the ship. What are its distance and bearing from the port?

9   Two planes leave an airport at the same time. One flies on a bearing of 330° at 800 km/h, while the other flies due east at 1200 km/h. Draw one diagram to show their positions and distances apart after:

   (a) $\frac{1}{4}$ hour;   (b) $\frac{1}{2}$ hour;   (c) $\frac{3}{4}$ hour;   (d) 1 hour.

10 From the top of a hill three church spires can be seen.
   The first, A, is 6 km away on a bearing of 330°; the second, B, is 4 km away on a bearing of 070°; and the third, C, is 2 km due south.
   Draw a plan of the positions of the spires using a scale of 1 cm to 1 km.

   (a) Measure the distances:

      (i) A to B,   (ii) B to C,   (iii) C to A.

   (b) What are the bearings of:

      (i) A from B,   (ii) B from C,   (iii) C from A?

Angle

4. RADAR

Many of you will have seen a rotating aerial on ships or at ports and aerodromes. These are *radar* aerials. Radar is used to 'see' objects that are far away or which are hidden in mist or cloud. A signal like a radio or light wave is sent out from the aerial. It bounces off any object it meets, is picked up again by the aerial and appears on a *radar screen* in the form of a bright spot of light. The actual screen is like a television set except that it has angles and concentric circles marked on it.

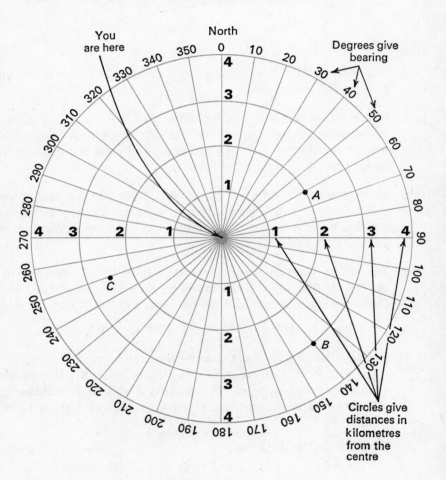

Imagine that you are the centre of the screen which represents a bird's-eye-view of the whole area around you. A point of light appearing on the screen will correspond to an object on the ground and you can immediately read off its distance and bearing. For example:

Radar

4. RADAR

This is an interesting introduction to polar coordinates. It also serves as a useful revision of the idea that an ordered pair of numbers can be used to describe the position of a point. If a discussion on this does arise, it might be worth illustrating the different systems of coordinates by a diagram such as Figure G.

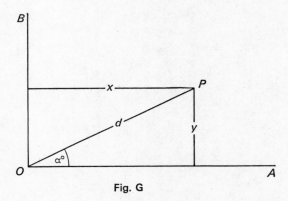

Fig. G

The Cartesian coordinates of $P(x, y)$ state the distances from the two fixed lines OA and OB. Polar coordinates (d, α) state a distance and an angle; the distance d of the point P from a fixed point O and the angle α (in degrees) which OP makes with some fixed line (in this case OA; in the case of the radar fixes, the North–South direction).

A master radar screen can be scratched on acetate sheet and used to great advantage on the overhead projector.

Angle

Exercise D

1   A (6, 030) ; B (3½, 060) ; C (7½, 105) ; D (4, 140) ;
    E (7, 140) ; F (3½, 217) ; G (6, 250) ; H (8, 290) ;
    I (4, 300) ; J (6, 345).

Radar

A is 2 km on a bearing of 060°;

B is 3 km on a bearing of 140°;

C is 2·5 km on a bearing of 250°.

We could write A as (2, 060) and remember that the first number stands for kilometres and the second number for degrees. This would save a lot of time and space if we had several positions to record.

How would you write the positions of B and C in this short way?

Exercise D

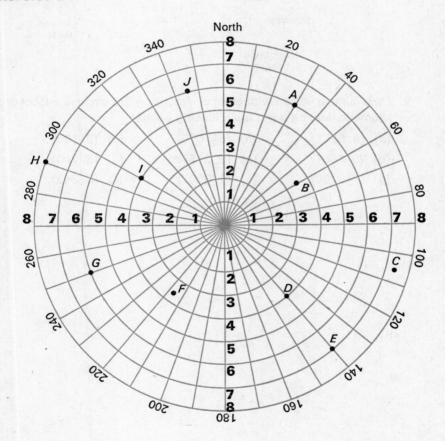

1 Use the shorthand method to give the distance and bearing of each of the points A to J on the radar screen above.

Angle

2 Trace this diagram of a harbour, and draw a radar screen centred on the Harbour master's office. Use concentric circles so that every centimetre represents 1 kilometre. Give the distance and bearing of all the points mentioned.

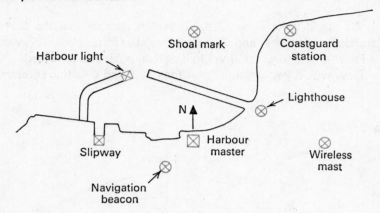

3 Draw a radar screen with eight circles, using 0·5 cm to represent each kilometre, and on it put the following points:

(a) (4, 025);   (b) ($6\frac{1}{2}$, 070);   (c) (2, 105);
(d) (8, 150);   (e) (8, 215);   (f) (5, 260);
(g) ($7\frac{1}{2}$, 260);   (h) (1, 300);   (i) (3, 343).

Radar

2   Shoal mark (3, 360). Coastguard station (4, 040).
    Lighthouse (2, 065). Wireless mast (3½, 090).
    Navigation beacon (1, 225). Slipway (2½, 270).
    Harbour light (2½, 315).

7. Relations

In a traditional course, it was assumed that pupils would understand what was meant by, for example, 'There is a relation between x and y, given by "$y = x^2$"'. Relations were always stated by means of graphs or formulae. In this course, we shall start a little further back and attempt to give a greater experience and awareness of relations before reaching that stage.

We also want to widen the concept of a relation in mathematics to extend beyond those that can be expressed in terms of formulae. For example, in this first year, i.e. in *Books A* and *B*, we have discussed relations between:

objects counted—the set of counting numbers;
positions in a plane—number pairs in the form of coordinates;
parts of diagrams—number pairs in the form of fractions;
binary numbers—various two-state systems;
numbers of sides of a regular polygon—interior angles of the polygon;
types of vehicle in a street—numbers indicating their frequency.

In this chapter we shall attempt to go from intuitive ideas about relations in the family sense to the first notion about relations in the mathematical sense, a notion which is usually expressed in terms of subsets of sets. We shall deal only with binary relations, that is relations of the form a is related to b, where a, b, are members of a given set.

1. FAMILY RELATIONSHIPS

We start with family relationships because we believe that these are the relationships with which the pupils are most familiar. In families, the words 'relation' and 'relative', indicating people, are synonymous and so we need the word 'relationship' to indicate the links between them. After the first section we shall use the word 'relation' to indicate the linking, the 'referring or comparing of two things one to another'.

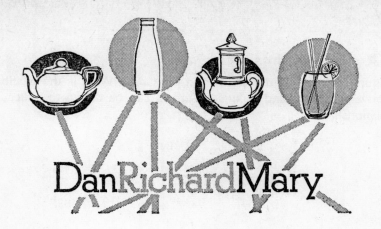

7. Relations

1. FAMILY RELATIONSHIPS

Figure 1 is a Family Tree. Why is it called a tree? This one tells you how a group of 15 people are related to one another. What does ' = ' mean? Does it make any difference if the first name is a man's or a woman's? What do the vertical lines mean? What do the horizontal lines indicate?

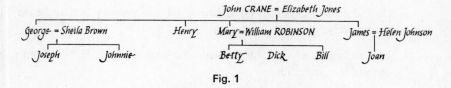

Fig. 1

Exercise A

1. How many children have John and Elizabeth?
2. How many grandchildren have they?
3. How many children call Henry 'Uncle'?
4. How many nephews has he?
5. What is Johnnie's surname?
6. Write down the full names of everyone who has a brother.
7. How many different relationships are represented in this family?

Relations

1.1 Arrow diagrams

Figure 2 shows all the children in the family. What is the relationship between Joseph and Johnnie? Arrows can be drawn to indicate this relationship.

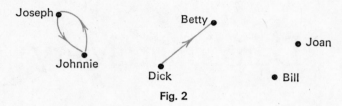

Fig. 2

Dick is the brother of Betty,
Joseph is the brother of Johnnie,
Johnnie is the brother of Joseph.

Copy Figure 2 and put in the other arrows that show 'is the brother of'.

Exercise B

1 In the family in Section 1, how many people have an uncle?

2 In a diagram like Figure 2, show all the members of the family that are necessary for the relationship

'is the uncle of'.

Complete the diagram by drawing all the arrows that indicate this relationship.

3 Draw diagrams with the sub-sets of people from Figure 1 and the arrows that indicate:
   (a) 'is the father of';
   (b) 'is the cousin of';
   (c) 'is the brother of'.

4 What do the arrows in Figure 3 mean? What extra arrows have to be drawn to show the relationship represented by '=' in Figure 1?

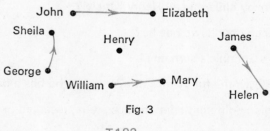

Fig. 3

Family relationships

Exercise A

1. 4.   2. 6.   3. 6.   4. 4.

5. Crane.

6. George Crane, Henry Crane, Mary Crane, James Crane, Joseph Crane, Johnnie Crane, Betty Robinson, Dick Robinson, Bill Robinson. We don't know about John Crane or his wife.

7. Wife,   Mother,   Daughter,   Grandmother,   Grand-daughter,
   Husband, Father,   Son,   Grandfather,   Grandson,
   Sister, Sister-in-law,
   Brother, Brother-in-law,
   Aunt,   Niece,
   Uncle,   Nephew,   Cousin.

   This list gives the 19 simple family relationships within the diagram. The diagram also indicates other relations, for example, 'is older than' between offspring of the same parents (the convention is that offspring are shown in order with the oldest on the left), 'is related by marriage to', 'is the only sister of'.

Exercise B

1. 6.

2.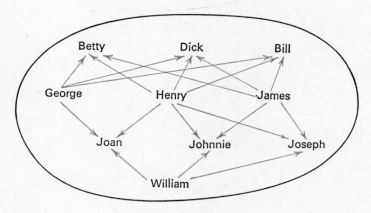

Fig. A

William Robinson would have to be removed from this set if the relationship 'is the uncle of' is taken to exclude uncles by marriage.

T183

Relations

3 (a)

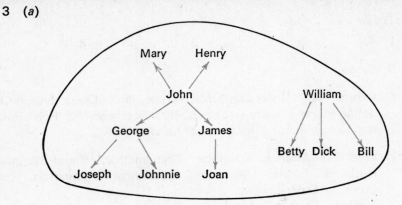

Fig. B

(b)

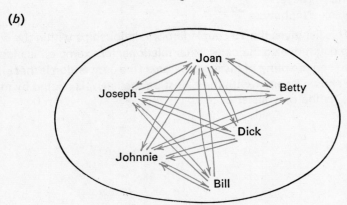

Fig. C

(c)

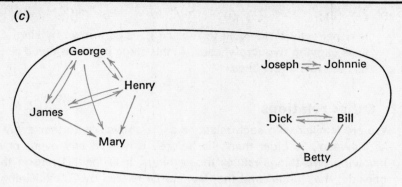

Fig. D

4 'is the husband of'. Arrows in the reverse direction to show the relation 'is the wife of'. Each pair is equivalent to 'is married to'.

Relations

5 (a) (ii);   (b) (i);   (c) (iii).

It is perfectly reasonable to replace pairs of arrows by single arrow lines carrying two arrowheads. At this stage it is considered clearer to use the two arrow lines.

1.2 Other relations

We are familiar with such relations as 'is above', 'is further away than', 'is between', 'is older than', 'is before'. However, *any* event or action involving two things relates those things. It is for this reason that we consider the 'judge' and the 'law' to be related by 'is explaining', and 'the picture' and 'John' to be related by 'is being painted by'. This point will be taken up again when we consider mappings.

Exercise C

3 Showing just the ages:

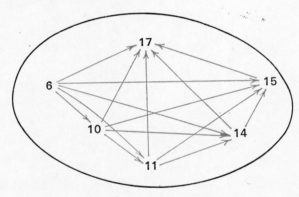

Fig. E

4 (a)   (b)

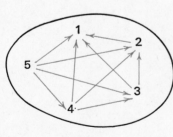

Fig. F

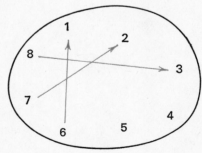

Fig. G

Family relationships

5 Are there (i) always, (ii) never, or (iii) sometimes two arrows between two people to show the relationship
    (a) 'is the father of';
    (b) 'is the cousin of';
    (c) 'is the brother of'?

1.2 Other relations

The moon is above the clouds.
The post office is further away than the grocer's.
The judge is explaining the law.
The picture is being painted by John.

These sentences indicate relations between pairs of things. The relations (we shall no longer use the word relationship) are
    'is above',
    'is further away than',
    'is explaining',
    'is being painted by'.

Exercise C

1 Write down three pairs of objects in the classroom which satisfy each of the following relations:
    (a) is in front of;   (b) is taller than;
    (c) is older than;   (d) is owned by;
    (e) is hidden by.

2 State at least two relations that hold between:
    (a) you and your neighbour;   (b) you and your desk;
    (c) you and your hands;   (d) an eagle and a sparrow;
    (e) a ship and the sea.

3 The following are the ages of the children of the family in Figure 1:
Joseph, 17; Johnnie, 14; Betty, 15; Dick, 11; Bill, 6; Joan, 10.
Make an arrow diagram with this set of children and the relation 'is younger than'.

4 Draw arrow diagrams to illustrate the relations:
    (a) 'is greater than' on members of the set $\{1, 2, 3, 4, 5\}$;
    (b) 'is 5 more than' on members of the set $\{1, 2, 3, 4, 5, 6, 7, 8\}$.

5 Draw an arrow diagram to illustrate the relations:
    (a) 'is north of' on members of a set of cities;
    (b) 'has more people than' on members of a set of countries;
    (c) 'is the same shape as' on members of a set of your text-books.

T187

Relations

1.3 Relations between sets

Make a list of, say, six boys or girls who sit near to you in the classroom, including yourself, and also a list of the different drinks they had yesterday. If a person had a particular drink, then we indicate this by drawing a line as in Figure 4.

For example, Dick had tea and lemonade.

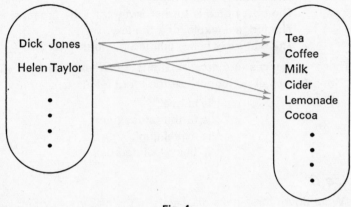

Fig. 4

Which was the most popular drink?
Who had the widest variety of drinks?
In this example, the members of two sets are linked by a relation.

In the same way, it is possible to have a relation between the members of two sets of numbers. For example, the set {2, 3, 5} and the set {3, 6, 10, 15} can have their members related by 'is a prime factor of'.

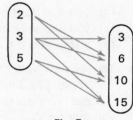

Fig. 5

Exercise D

1 List a set of family pets and another set of people you know. Use an arrow diagram to illustrate the relation 'is owned by'.

2 List a set of people you know and a set of television programmes. Draw arrows to illustrate the relation 'watched the programme'. Which was the favourite programme? Who watched the largest number of programmes?

Family relationships

1.3 Relations between sets

In the given family, Henry 'is uncle of' Joseph. This is a particular instance of a relation. If we wanted to define the complete relation 'is uncle of' on this family, we should have to state or show all possible instances as in Figure A. In that case, we may state that the relation holds between the elements of the two sets,

$$\left\{\begin{array}{l}\text{George, Henry,}\\ \text{William, James}\end{array}\right\} \text{ and } \left\{\begin{array}{l}\text{Joseph, Johnnie, Betty,}\\ \text{Dick, Bill and Joan.}\end{array}\right\}$$

For short, we say that there is a relation between the sets.

Relations

Exercise D

5

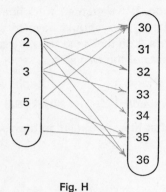

Fig. H

31 has no arrows joined to it because it is a prime number.

6

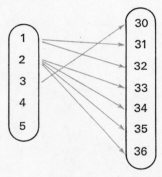

Fig. I

2. MAPPINGS

The concept of a relation is an essentially static one; it involves two sets whose elements are in some way associated. But we saw in Section 1.2 that actions can establish a relation between two elements. They can do the same between sets of elements. For example, transformation geometry is the study of certain kinds of relation between sets of points. To begin with, pupils establish these relations by seeing how one set can be moved onto another (using, perhaps, tracing paper or folding). Motions such as rotation or translation are initially considered as actions which generate one set of points from another.

Family relationships

3   Make a list of Junior schools in your area and some of the people in your class. Use an arrow diagram to show the relation 'used to be the school of'. Are there any schools from which especially large numbers have come? Why is this?

4   Draw arrow diagrams to represent the relations

   (a) 'is the height of';
   (b) 'is the age of';
   (c) 'is the number of brothers of';

   between suitable sets of numbers and some members of your class.

5   Draw a diagram to represent the relation 'is a factor of' between the members of {2, 3, 5, 7} and the members of {30, 31, 32, 33, 34, 35, 36}. Which number had no arrows joined to it? Why?

6   Draw a relation diagram between the members of {1, 2, 3, 4, 5} and the members of {30, 31, 32, 33, 34, 35, 36} to illustrate the relation 'is the number of prime factors of'.

2. MAPPINGS

If the members of the set {1, 2, 3, 4, 5} are each doubled, a new set is formed, {2, 4, 6, 8, 10}. Doubling transforms members of one set into members of the other. The transformation can be illustrated as an arrow diagram:

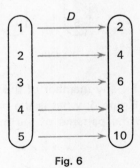

Fig. 6

We say that the members of one set are *mapped* onto the members of the other. Under the mapping, 4, for example, is called the *image* of 2. A mapping is a special kind of relation in which each member of one set is related to exactly one member of the set of images. (Figure 4 does not illustrate a mapping.)

Relations

We can imagine a mapping machine.

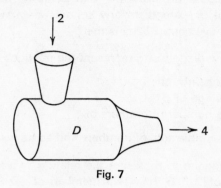

Fig. 7

When 2 is put into the machine, it is transformed into 4. The machine will double any number put into it. 4 is called the *image* of 2 under the mapping *D*. Notice that the machine can only produce one answer.

The following are also mapping diagrams. What is the machine doing in each case?

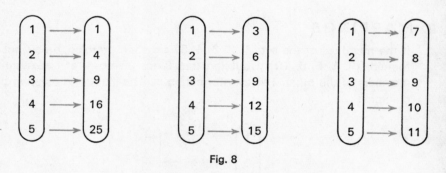

Fig. 8

If we let *x* stand for any member of the set {1, 2, 3, 4, 5} under the mapping shown in Figure 6, what is the corresponding member of the other set? We show the patterns of the mappings by using letters for numbers like this:

$$x \to 2 \times x,$$

which we usually write as $x \to 2x$.

The patterns for the mappings in Figure 8 are:

$$x \to x^2,$$
$$x \to 3x,$$
$$x \to x + 6.$$

86 T192

Mappings

This idea of an action which connects sets is very valuable in the classroom and it is here extended from points to numbers by the invention of the 'mapping machine'. However, notice that we use the word 'mapping' only when each member of the first set has a unique image in the second set.

Relations

Exercise E

1 The image sets under the mappings are:

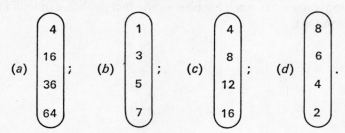

2 (a) $x \to x+1$; (b) $x \to 5x$; (c) $x \to x+20$;
   (d) $x \to 100x$; (e) $x \to 1/x$; (f) $x \to 5-x$.

3 (i)   (a) 1;   (b) 81;   (c) 25;   (d) 0.
   (ii) (a) 13; (b) 21;   (c) 17;   (d) 12.
   (iii) (a) 49; (b) 41;   (c) 45;   (d) 50.

4 (i)   (a) 6;   (b) 7;   (c) 4;   (d) 5.
   (ii) (a) 24; (b) 37;   (c) 4;   (d) 13.
   (iii) (a) 14; (b) 1;   (c) 34;   (d) 25.

5 90° anticlockwise.
   (a) $A \to C$; $B \to D$; $C \to A$; $D \to B$;
   (b) $A \to B$; $B \to C$; $C \to D$; $D \to A$.

Mappings

Exercise E

1. Draw mapping diagrams to show {2, 4, 6, 8} under the following mappings:
   (a) $x \to x^2$;
   (b) $x \to x-1$;
   (c) $x \to 2x$;
   (d) $x \to 10-x$.

2. Express the following mappings in the form '$x \to ?$':

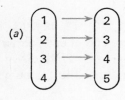 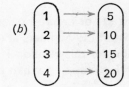 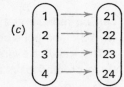

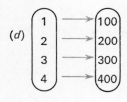 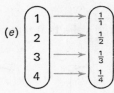 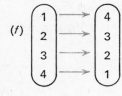

Fig. 9

3. What are the images of (a) 1, (b) 9, (c) 5, and (d) 0 under the mappings
   (i) $x \to x^2$;
   (ii) $x \to x+12$;
   (iii) $x \to 50-x$?

4. What numbers have the following images under the mappings of Question 3:
   (a) 36;
   (b) 49;
   (c) 16;
   (d) 25?

5. A square has rotational symmetry of order 4. The square ABCD is rotated about its centre so that

   $A \to D$; $B \to A$; $C \to B$; $D \to C$.

   Through what angle has the square been rotated? On to what points are A, B, C and D mapped after an anticlockwise rotation of (a) a half-turn, and (b) a three-quarter-turn from its original position?

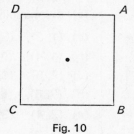

Fig. 10

Relations

3. ORDERED PAIRS

The relation

Fig. 11

can be represented as a set of ordered pairs, (1, 10), (2, 11), (3, 12), (4, 13). The pairs consist of the numbers at either end of each arrow.

The general expression for this relation is $x \to x+9$. Write down the three ordered pairs for this mapping when it is applied to the numbers 5, 6, and 7.

Exercise F

1   Write down the ordered pairs that represent the relations:

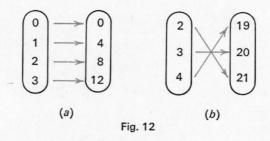

(a)   (b)

Fig. 12

2   Write down the ordered pairs that represent the mapping $x \to 4x$ applied to the set {10, 20, 30, 40}.

3   The following sets of ordered pairs are relations. State them in the form $x \to$? giving the set to which they are applied. For example, in (a), the mapping $x \to 3x$ is applied to the set {1, 3, 5, 7}.

  (a) (1, 3), (3, 9), (5, 15), (7, 21);
  (b) (1, 4), (3, 6), (5, 8), (7, 10);
  (c) (2, 5), (3, 6), (4, 7), (5, 8);
  (d) (7, 5), (6, 4), (5, 3), (4, 2);
  (e) (2, 4), (3, 6), (5, 10), (7, 14);
  (f) (1, 9), (3, 7), (6, 4), (10, 0).

T196

Ordered pairs

3. ORDERED PAIRS

Any relation between two finite sets of numbers can be shown by drawing an arrow between the associated individual numbers or by writing down the associated numbers as ordered pairs. The advantage of the latter scheme is that these pairs can be taken as coordinates and the relation illustrated as a graph. This is the work of the chapter in *Book C* called 'From Relation to Graph'. The present section is a link between the two chapters. It provides a first glance at the idea of relation as a set of ordered pairs and so it is suggested that not too much emphasis need be placed upon it at this stage. It will be taken up again later.

Exercise F

1. (a) (0, 0), (1, 4), (2, 8), (3, 12);
   (b) (2, 21), (3, 20), (4, 19).

2. (10, 40), (20, 80), (30, 120), (40, 160).

3. (a) $x \to 3x$ applied to the set $\{1, 3, 5, 7\}$;
   (b) $x \to x+3$ applied to the set $\{1, 3, 5, 7\}$;
   (c) $x \to x+3$ applied to the set $\{2, 3, 4, 5\}$;
   (d) $x \to x-2$ applied to the set $\{7, 6, 5, 4\}$;
   (e) $x \to 2x$ applied to the set $\{2, 3, 5, 7\}$;
   (f) $x \to 10-x$ applied to the set $\{1, 3, 6, 10\}$.

Relations

4 See Figure J. Each set of points lies on a straight line. The lines for (*a*) and (*e*) would pass through the origin.

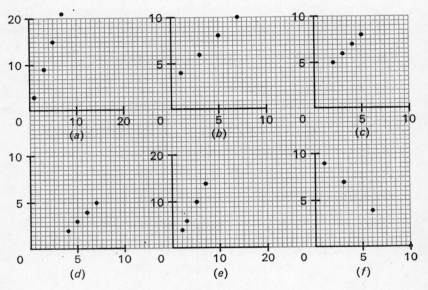

Fig. J

5 (*a*) (1, 4), (2, 8), (3, 12), (4, 16), (5, 20);
   (*b*) (1, 5), (2, 6), (3, 7), (4, 8), (5, 9);
   (*c*) (1, 1), (2, ½), (3, ⅓), (4, ¼), (5, ⅕);
   (*d*) (1, 5), (2, 4), (3, 3), (4, 2), (5, 1).

Only (*a*) is a straight line that passes through the origin. (*c*) is not a straight line.

Ordered pairs

4 The fact that a relation is a set of ordered pairs suggests that the relation can be shown on graph paper, using the first member of the pair to represent the x-coordinate and the second number to represent the y-coordinate.

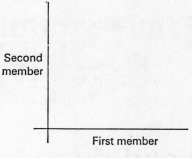

Fig. 13

Plot the ordered pairs of the six parts of Question 3. What do you notice about each set of plotted points? If lines were drawn through these plotted points, would any of them pass through the origin?

5 Work out the ordered pairs that show the following mappings applied to the set {1, 2, 3, 4, 5}. Plot the ordered pairs you obtain. Which sets lie on a straight line through the origin? Which do not lie on a straight line?

(a) $x \to 4x$; (b) $x \to 4+x$; (c) $x \to \dfrac{1}{x}$; (d) $x \to 6-x$.

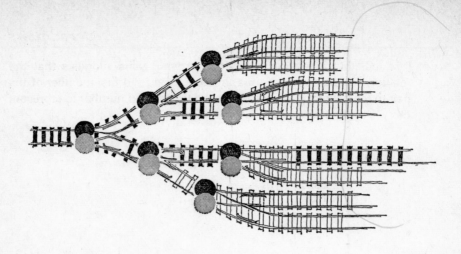

8. Binary and duodecimal bases

1. REVISION

In Book A we saw how we could do arithmetic in bases other than ten, and before going on with new work, some revision would be useful.

Exercise A

1 Copy and complete this cross number. Clues are given in base ten, give the answers in base five.

Across	Down
1. 8	1. 49
3. 39	2. 19
6. 124	4. 66
7. 15	5. 20
8. 21	9. 87
9. 18	10. 103
11. 12	11. 11
12. 90	13. 19
14. 38	
15. 23	

Write your answers to the following questions in the given base.

8. Binary and duodecimal bases

1. REVISION

This chapter is a continuation of the work started in Chapter 4 of *Book A*, and begins by revising that chapter. If more revision is needed, there could be a discussion and construction of Napier's bones. They form a simple mechanical calculating device for multiplication (Napier also invented the logarithm). The following examples show the use of them in base five.

A multiplying device for base five (Napier's bones)

Cut six pieces of card and number them as below:

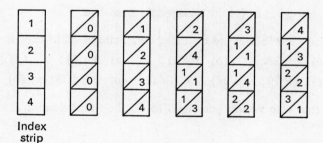

Fig. A

Pupils should be aware of how the 'bones' are constructed and not regard them as 'magic'.

Binary and duodecimal bases

To use them, take the index strip together with the other 'bones' required as in these examples:

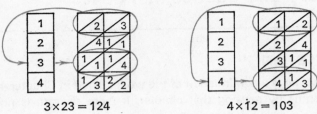

$3 \times 23 = 124$ $4 \times 12 = 103$

Fig. B

Notice that when numbers from adjacent 'bones' are next to each other they have to be added. It is useful to show how this process compares with long multiplication using paper and pencil.

Exercise *(Answers are printed on opposite page.)*

1. Make a set of Napier's bones for base five.

2. Use your 'bones' to do these calculations, all in base five.
   (a) 2×23;   (b) 3×31;   (c) 3×41;   (d) 4×32;
   (e) 3×123;   (f) 2×104;   (g) 3×403;   (h) 4×1234.

3. What else would you need to do the multiplication 3×221 in base five?

4. Design a set of bones to multiply in base eight.

Revision

Answers

2 (a) 101;   (b) 143;   (c) 223;   (d) 233;
   (e) 424;   (f) 213;   (g) 2214;   (h) 11101.

3 An extra 'bone' headed '2' would be needed.

4

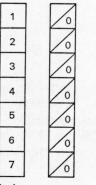

Fig. C

Exercise A

1

Fig. D

Binary and duodecimal bases

2	42.	3	124.
4	111.	5	211.
6	24.	7	13.
8	223.	9	1002.
10	1744.	11	4760.
12	13.	13	5.
14	6.		

2. BASE TWELVE

It is most important that at least one base greater than ten be studied in order to emphasize the need for extra symbols in the units column.

For a fairly detailed treatment of the Duodecimal System, see *An Introduction to Number Scales and Computers,* Chapter 2, by F. J. Budden. This gives details of the proposals made by the Duodecimal Societies of Britain and America.

Revision

2 $23_{five} + 14_{five}$ 3 $45_{six} + 35_{six}$

4 $13_{four} + 32_{four}$ 5 $133_{five} + 23_{five}$

6 $43_{six} - 15_{six}$ 7 $32_{five} - 14_{five}$

8 $45_{six} \times 3_{six}$ 9 $23_{four} \times 12_{four}$

10 $123_{eight} \times 14_{eight}$ 11 $476_{nine} \times 10_{nine}$

12 $31_{five} \div 2_{five}$ 13 $32_{six} \div 4_{six}$

14 $60_{eight} \div 10_{eight}$

2. BASE TWELVE

We can make a spike abacus with spikes as tall as we like. Examine this one.

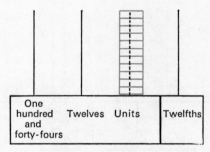

Eleven reels have been added to the units spike; if we try to add one more we shall have to change it and the eleven reels already on the units spike for a single reel on the next spike. This will stand for a group of twelve. We shall be counting in twelves.

The next three situations are easily represented in number symbols. But what about the last two?

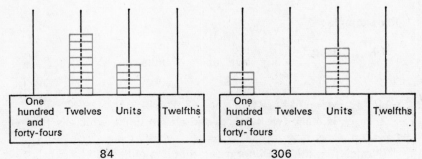

Binary and duodecimal bases

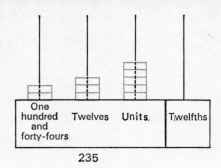

235

We want a single number symbol to represent the pile of reels on each spike, but we do not have a single symbol for the number ten or eleven.

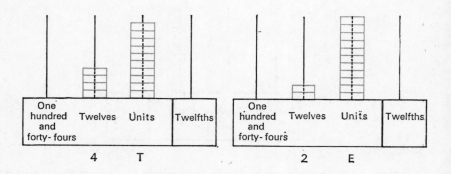

We shall have to invent new symbols for ten and eleven, and the simplest seem to be *T* and *E*.

Counting in base twelve will then go:

0, 1, 2, 3, 4, 5, 6, 7, 8, 9, *T*, *E*, 10, 11, 12, 13, and so on.

For example:

14_{twelve} is really

twelves	units
1	4

which is sixteen in base ten.

$2E_{twelve}$ is really two twelves and eleven units, which is thirty-five in base ten.

(Notice that this does *not* mean that 2 is multiplied by *E*.)

Base twelve

It is profitable here to discuss with the pupils whether twelve is better than ten as a number base. Obvious points in favour are:
(i) A dozen is quite a common unit of retail trade.
(ii) Twelve has more factors than ten; with base twelve we have:

$\frac{1}{2}$ of twelve is six
$\frac{1}{4}$ of twelve is three
$\frac{3}{4}$ of twelve is nine
$\frac{1}{3}$ of twelve is four
$\frac{2}{3}$ of twelve is eight
$\frac{1}{6}$ of twelve is two
$\frac{5}{6}$ of twelve is ten

But with base ten we can have only:

$\frac{1}{2}$ of ten is 5
$\frac{1}{5}$ of ten is 2
$\frac{2}{5}$ of ten is 4
$\frac{3}{5}$ of ten is 6
$\frac{4}{5}$ of ten is 8

Be sure not to overlook the extra interest which can be added with calculations such as 'five toes' and 'two tots'; see Exercise B, Questions 21, 22, etc.

Binary and duodecimal bases

Exercise B

1. (a) 13; (b) 18; (c) 19; (d) 26; (e) 1T;
   (f) 30; (g) E; (h) 4T; (i) 3E; (j) 100.

2. (a) 13; (b) 48; (c) 10; (d) 23; (e) 61;
   (f) 72; (g) 46; (h) 120; (i) 141, (j) 131.

3.

+	0	1	2	3	4	5	6	7	8	9	T	E
0	0	1	2	3	4	5	6	7	8	9	T	E
1	1	2	3	4	5	6	7	8	9	T	E	10
2	2	3	4	5	6	7	8	9	T	E	10	11
3	3	4	5	6	7	8	9	T	E	10	11	12
4	4	5	6	7	8	9	T	E	10	11	12	13
5	5	6	7	8	9	T	E	10	11	12	13	14
6	6	7	8	9	T	E	10	11	12	13	14	15
7	7	8	9	T	E	10	11	12	13	14	15	16
8	8	9	T	E	10	11	12	13	14	15	16	17
9	9	T	E	10	11	12	13	14	15	16	17	18
T	T	E	10	11	12	13	14	15	16	17	18	19
E	E	10	11	12	13	14	15	16	17	18	19	1T

Base twelve

Exercise B

1 Copy and complete this table.

	Base ten	Base twelve
a	15	
b	20	
c	21	
d	30	
e	22	
f	36	
g	11	
h	58	
i	47	
j	144	

2 Copy and complete this table.

	Base ten	Base twelve
a		11
b		40
c		T
d		1E
e		51
f		60
g		3T
h		T0
i		E9
		TE

3 Copy and complete this addition table in base twelve.

+	0	1	2	3	4	5	6	7	8	9	T	E
0												
1												
2												
3												
4												
5												
6												
7												
8												
9												
T												
E												

Do not forget to look for patterns.

T209

Binary and duodecimal bases

4 Make out and complete a multiplication square in base twelve.

The next questions are in base twelve, give your answers in base twelve also.

5	57+28		6	39+28
7	36+46		8	67+55
9	79+1*T*		10	*T*0+1*T*
11	*T*0*T*+11		12	147+307
13	36−19		14	65−28
15	77−39		16	81−22
17	46×2		18	1*T*×3
19	12×*T*		20	*TE*×10

21   *T*1*E*
   + *T*0*E*
―――――

22   *T*0*E*
  ×   5
―――――

23 Try to explain the difference between working in the old units, feet and inches, and working in base twelve. Make up some examples to illustrate your points.

2.1 Bases greater than twelve

We have seen that we can make the abacus spikes as long as we like, so that we can work in bases greater than ten, and we took base twelve as an example. Even greater bases would be possible if we invented more number symbols.

For example, if we used base fourteen then we would need symbols for ten, eleven, twelve and thirteen. We could use *T, J, Q, K*. Counting in base fourteen would then go:

0, 1, 2, 3, 4, 5, 6, 7, 8, 9, *T, J, Q, K*, 10, 11, 12, 13, 14, 15, 16, 17, 18, 19, 1*T*, 1*J*, 1*Q*, 1*K*, 20, and so on.

Exercise C

1 (*a*) Invent the extra number symbols needed to work in base sixteen.
  (*b*) Write out the numbers from nought to thirty-two in base sixteen, using the symbols you have invented.

Base twelve

4

x	0	1	2	3	4	5	6	7	8	9	T	E
0	0	0	0	0	0	0	0	0	0	0	0	0
1	0	1	2	3	4	5	6	7	8	9	T	E
2	0	2	4	6	8	T	10	12	14	16	18	1T
3	0	3	6	9	10	13	16	19	20	23	26	29
4	0	4	8	10	14	18	20	24	28	30	34	38
5	0	5	T	13	18	21	26	2E	34	39	42	47
6	0	6	10	16	20	26	30	36	40	46	50	56
7	0	7	12	19	24	2E	36	41	48	53	5T	65
8	0	8	14	20	28	34	40	48	54	60	68	74
9	0	9	16	23	30	39	46	53	60	69	76	83
T	0	T	18	26	34	42	50	5T	68	76	84	92
E	0	E	1T	29	38	47	56	65	74	83	92	T1

In both Questions 3 and 4, the spotting of patterns will speed completion of the tables. Since addition and multiplication are both commutative, we could well do with only half of each table. This is a useful point for discussion, although the word commutative need not be mentioned.

5 83.   6 65.   7 80.   8 100.   9 97.
10 *ET*.   11 *T1E*.   12 452.   13 19.   14 39.
15 *3T*.   16 *5E*.   17 90.   18 56.   19 *E8*.
20 *TO*.   21 *182T*.   22 4247.

2.1 Bases greater than twelve

In discussing bases greater than twelve, attempts will have to be made to make up more symbols. Such attempts cause pupils to realize that it is not such a simple task as it at first appears, particularly if the symbols are to remain simple and easy to remember. This might be an appropriate stage to discuss the development of our present number symbols, 0 to 9.

Binary and duodecimal bases

Exercise C

1. (*a*) A possible answer 0, 1, 2, 3, 4, 5, 6, 7, 8, 9, *T, J, Q, K, A, X*.

   (*b*) As above, then 10, 11, 12, 13, 14, 15, 16, 17, 18, 19, 1*T*, 1*J*, 1*Q*, 1*K*, 1*A*, 1*X*, 20.

2. (*a*) 6*Q*;   (*b*) 7*T*;   (*c*) 6*J*;   (*d*) 9*X*;   (*e*) 60;
   (*f*) 51;   (*g*) 61;   (*h*) 70;   (*i*) 75.

3. 16 lb do not make a further unit of weight and so on.

3. BASE TWO

This can be approached by first considering that there is no upper limit to the number base—the spikes on the abacus can be made as long as we like. Next consider how short the spikes can become—counting is still possible with spikes which are only long enough to hold one reel.

Base twelve

2 All these questions are in base sixteen and answers are also required in base sixteen.

  (a) 49+23;   (b) 36+44;
  (c) 25+46;   (d) 48+57;
  (e) 38+28;   (f) 19+38;
  (g) 18+49;   (h) 38×2;
  (i) 27×3.

3 What is the difference between working in base sixteen and working in the old units, pounds and ounces? Try and explain this in your own words and illustrate your answer with examples.

3. BASE TWO

The shortest spike we can put on the abacus is one that will take only one reel. Any attempt to put on two means we have to change spikes.

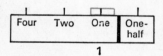

1 represents one

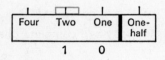

10 represents two

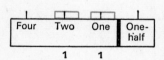

11 represents three

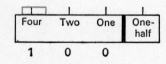

100 represents four

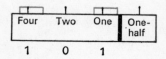

101 represents five

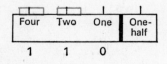

110 represents six

Binary and duodecimal bases

Exercise D

1. Imagine you had a very long abacus like this one and only one reel would go on each spike.

   Make a copy of this abacus and write the name given to each spike in the space beneath it.

2. What are the only digits that can be used in this base?

3. Copy and complete this table.

	Base ten	Base two
a	3	
b	5	
c	10	
d	12	
e	15	
f	20	
g	25	
h	26	
i	30	
j	32	

4. Copy and complete this table.

	Base ten	Base two
a		10
b		100
c		111
d		1001
e		1100
f		1110
g		1000
h		10001
i		11111
j		100000

5. Copy and complete these addition and multiplication tables in base two.

   (a)

+	0	1
0		
1		

   (b)

×	0	1
0		
1		

Base two

Exercise D

1

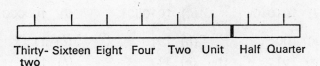

Fig. E

2 0 and 1.

3 (a) 11;   (b) 101;   (c) 1010;   (d) 1100;   (e) 1111;
(f) 10100;   (g) 11001;   (h) 11010;   (i) 11110;   (j) 100000.

4 (a) 2;   (b) 4;   (c) 7;   (d) 9;   (e) 12;
(f) 14;   (g) 8;   (h) 17;   (i) 31;   (j) 32.

5 (a)

+	0	1
0	0	1
1	1	10

(b)

×	0	1
0	0	0
1	0	1

Binary and duodecimal bases

6 (a) 1111;   (b) 110;   (c) 1011;
   (d) 1000;   (e) 100111;   (f) 100010;
   (g) 101110; (h) 10000;   (i) 10010;
   (j) 101010; (k) 101000;   (l) 1000001.

7 (a) 100;   (b) 10;   (c) 11;
   (d) 101;   (e) 11;   (f) 111;
   (g) 1110;   (h) 101;   (i) 10100.

8 (a) 1010;   (b) 10101;
   (c) 100100;   (d) 1000001;
   (e) 10000100;   (f) 1111110.

Base two

6 All these additions are in base two, give answers also in base two.

(a) 1010
   + 101
   ─────

(b) 101
   + 1
   ─────

(c) 110
   + 101
   ─────

(d) 111
   + 1
   ─────

(e) 11100
   + 1011
   ─────

(f) 10101
   + 1101
   ─────

(g) 100001
   + 1101
   ─────

(h) 1111
   + 1
   ─────

(i) 1111
   + 11
   ─────

(j) 10101
   + 10101
   ─────

(k) 11001
   + 1111
   ─────

(l) 110011
   + 1110
   ─────

7 All these subtractions are in base two, give answers also in base two.

(a) 111
   − 11
   ─────

(b) 101
   − 11
   ─────

(c) 100
   − 1
   ─────

(d) 110
   − 1
   ─────

(e) 110
   − 11
   ─────

(f) 1010
   − 11
   ─────

(g) 10111
   − 1001
   ─────

(h) 11100
   − 10111
   ─────

(i) 110011
   − 11111
   ─────

8 All these multiplications are in base two, give your answers in base two.

(a) 101 × 10;

(b) 111 × 11;

(c) 1001 × 100;

(d) 1101 × 101;

(e) 10110 × 110;

(f) 10010 × 111.

T217

Binary and duodecimal bases

9

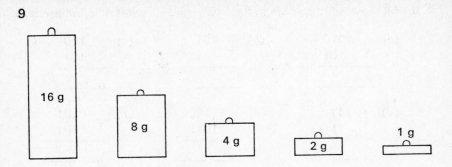

If you had this set of weights, list how you would find the weights of the quantities in the right-hand column.

Two examples have already been done for you.

16 g	8 g	4 g	2 g	1 g	Grammes
					1
					2
					3
					4
		1	0	1	5
					6
					24
					25
					26
1	1	0	1	1	27
					28
					29
					30
					31

3.1 The human computer

To make this, five (or more) pupils should stand out at the front and face the class, with both hands at their sides. These people form the 'calculating unit'. One other member of the class is needed as the 'input device'.

Numbers are fed into the computer by the input pupil, who makes a signal such as a rap on the desk or a single hand-clap. The computer counts every pulse which is fed into it in base two.

A alters at every pulse, 'If she is down she goes up and if she is up she goes down'.

Base two

9	1	10001	This question can be extended to examine the set of weights needed if weights can be placed in one or both scale pans. These will be found to be 1 g, 3 g, 9 g, 27 g, ..., the place values of base three.
	10	10010	
	11	10011	
	100	10100	
	101	10101	
	110	10110	
	111	10111	
	1000	11000	
	1001	11001	
	1010	11010	
	1011	11011	
	1100	11100	
	1101	11101	
	1110	11110	
	1111	11111	
	10000		

Binary and duodecimal bases

3.1 The human computer

This is the time to emphasize the idea of a 'two state system'. With base two, it is either one state or the other, either yes or no, up or down, left or right, on or off, and it is this feature of the base which makes it so applicable to electrical circuits. It is often possible to display a few numbers in base two with the ordinary light switches in a classroom.

Base two

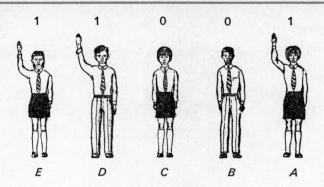

B alters at every second pulse.
C alters at every fourth pulse.
D alters at every eighth pulse.
E alters at every sixteenth pulse.

The computer illustrated above will count up to thirty-one, but more people can be added if required. It is best to start with just three people and keep adding extra people as everyone gets used to the idea.

3.2 Electronic computers

Base two is often called the 'Binary System' or the 'Binary Code' and it has become important because of its use in computers. Electronic computers depend on the flow of electric current.

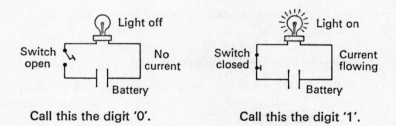

Call this the digit '0'.   Call this the digit '1'.

A row of light bulbs can be set up to represent numbers in the binary system; if the light is on call it '1' and if it is off, call it '0'.

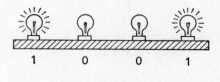

Binary and duodecimal bases

Exercise E

1 Write down the binary numbers represented by these lights and convert your answers to base ten.

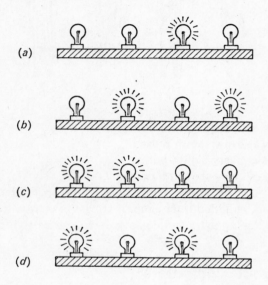

2 Illustrate these base ten numbers by lights in base two.

(a) 3; (b) 6; (c) 11; (d) 15.

3 What is the largest number that can be represented by 4 lights? How many lights would you need to represent one hundred?

3.3 DIBS (device to illustrate the binary system)

Here is the circuit for a simple device you could make for yourself.
   Materials: 4 bulb holders, 4 bulbs, 4 on-off switches, a battery, some wire and a base board.

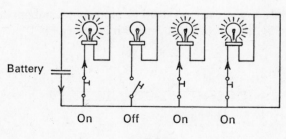

Base two

Exercise E

1. (a) 10_{two}, 2_{ten};　　(b) 101_{two}, 5_{ten};
   (c) 1100_{two}, 12_{ten};　　(d) 1010_{two}, 10_{ten};

2. (a) 11;　　(b) 110;　　(c) 1011;　　(d) 1111.

3. Fifteen. One hundred needs seven lights.

3.3 and 3.2 D.I.B.S. and Electronic computers

D.I.B.S. is not electronic, it is simply electric. It is really only four separate lights with no carry circuit, and is well within the capabilities of even the most unpractical pupil.

A circuit capable of carrying is much more complicated and details can be found in *Mathematical Models*, by Cundy and Rollett, or in *Mathematics Teaching No. 28*.

Figure F shows a simple circuit which will carry.

P.A.C.A. (*Put And Carry Adder*)

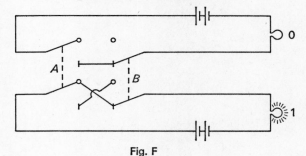

Fig. F

A and *B* are each really two switches linked together as shown by the line of dashes. There is no need for expensive and proper equipment to make this; a very successful model has been seen, made from wire and paper clips mounted on a piece of cardboard.

This device will illustrate:

$$0+0 = 0,$$
$$0+1 = 1,$$
$$1+0 = 1,$$
$$1+1 = 10.$$

Binary and duodecimal bases

Details of many more advanced models can be found in *We Built our own Computers*, an S.M.P. Handbook. Modern electronic construction kits are on sale in toy and model shops which will make transistorized models, but they are expensive.

Details of a simple mechanical device can be found in *Experiments in Mathematics*, Pearcy and Lewis, Longmans.

Exercise F

1   (*a*) 1, 24, 21, 2;   (*b*) 4, 8, 14, 19.

2   Thirty-one.   3   Mathematics.

Base two

You do not have to restrict yourself to just four lights; more can be added if desired.

The circuit on p. 100 will illustrate numbers in the binary scale up to fifteen, but it will not add, carry or do any other arithmetical processes. Circuits which will do this are more complicated and depend on special switches, relays and transistors.

3.4 Computer tape

A representation of base two numbers can easily be made by punching holes in a piece of paper tape. This is how numbers are fed into some computers. There are different ways of doing this and a simplified way is dealt with here.

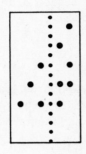

A hole stands for the digit one.
No hole stands for zero.

Exercise F

1 Write down the numbers in base two represented by these tapes. Convert your answers to base ten. Ignore the small holes, they are just to drive the tape through the machine.

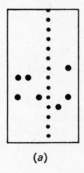   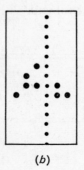

(a)   (b)

2 What is the largest number you can represent with this five-hole tape?

Binary and duodecimal bases

3 Words can also be put on to computer tape by first numbering the letters of the alphabet from one to twenty-six in base two. What does this tape say?

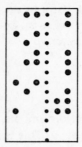

4 Make a drawing of a piece of paper tape which has your name on it in a pattern of holes as suggested in Question 3, and stick it on the front of your mathematics exercise book.

4. BASE TWO FRACTIONS

Going back to the idea of a base two spike abacus, we remember that the spare spikes stood for fractions or parts of units. We used a 'point' to show where the whole numbers ended and the fractions began.

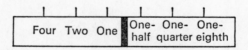

Exercise G

1 Copy the table and put the answers in. The first three have already been done for you.

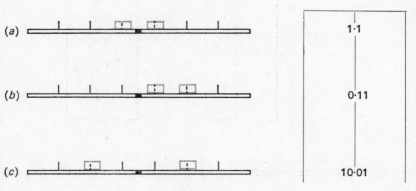

102   T 226

Base two fractions

Punched cards

As well as punched paper tape, punched cards can be made to record binary numbers.

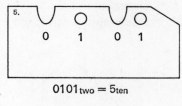

$0101_{two} = 5_{ten}$

Fig. G

Make a set of sixteen cards and number them from zero to fifteen. Punch four holes in the top of each one and cut a corner off. Cut a slot for zero and leave the hole to represent one as in Figure G. It is worth discussing (i) whether it would make any difference if the slot was meant to represent one and the hole to represent zero; (ii) why the corner is cut off. (Notice that if the card in Figure G were turned over it would represent 1010_{two}.)

The following notes show how the cards can be used.

1. *To select any card*

Take as example, card number five. First of all represent in the binary scale

$$5_{ten} = 0101_{two}$$

Take a nail or knitting needle and put it through the units hole of the pack. Shake so that all the cards with a slot in this position drop out and retain those still on the needle. Withdraw the needle and place through the two's hole, shake and retain all those which drop out. Put the needle through the four's hole and retain those on the needle. Put the needle through the eight's hole and retain those which drop out. This should be just one card, the one with 'five' on it.

2. *To place the cards in order*

First of all shuffle and make sure that the cards are mixed up. Put the needle through the units hole, lift out all those retained and place them *behind* the others. Repeat for the two's hole, then the four's hole and finally the eight's hole. The cards should now be in order from zero to fifteen.

Naturally, these processes can be greatly speeded up mechanically and this idea forms the basis of punched card systems.

Binary and duodecimal bases

Railway marshalling yards

An example of a two-state system can be found in the shunting of trucks in a marshalling yard; at each junction we can take 0 and 1 to mean either right or left fork. A truck can then have its destination converted into a binary code and this can be programmed to work the points automatically.

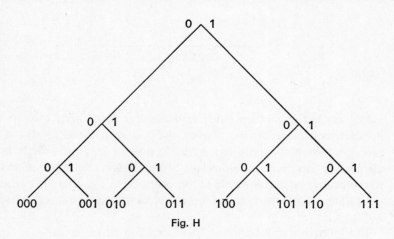

Fig. H

It is useful to show the class this example, since the diagram above is really a 'tree diagram' which will be encountered again in work on probability.

4. BASE TWO FRACTIONS

Exercise G

1. (d) 11.1_{two}, $3\frac{1}{2}_{ten}$;   (e) 1.11_{two}, $1\frac{3}{4}_{ten}$;
   (f) 11.001_{two}, $3\frac{1}{8}_{ten}$;   (g) 0.011_{two}, $\frac{3}{8}_{ten}$;
   (h) 100.101_{two}, $4\frac{5}{8}_{ten}$.

2. (a) 100·0;   (b) 1001·101;   (c) 1011·0000;
   (d) 110·000;   (e) 10·10;   (f) 11·110;
   (g) 1·110;   (h) 1·010;   (i) 10·1;
   (j) 1001·0;   (k) 1101·1.

3. (a) 11000·01;   (b) 1010·1001.

Base two fractions

(d)

(e)

(f)

(g)

(h)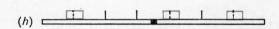

2 All these are in base two, give answers also in base two.

(a) 10·1 + 1·1

(b) 110·011 + 11·01

(c) 111·0101 + 11·1011

(d) 10·001 + 11·111

(e) 11·11 − 1·01

(f) 101·101 − 1·111

(g) 10·011 − ·101

(h) 10·101 − 1·011

(i) 1·01 × 10

(j) 11·0 × 11

(k) 11·011 × 100

3 These additions are in base two, give answers also in base two.

(a) 1110·01
     101·11
   + 100·01

(b) 101·101
     11·11
   + 1·0011

Binary and duodecimal bases

Exercise H (Miscellaneous)

1 Copy and complete these number scales for addition in base two.

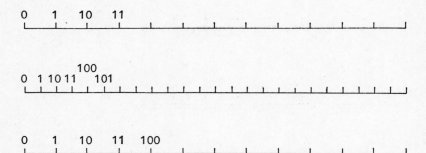

2 Draw a diagram to show how the scales are used to add 1001 to 11.

3 Copy and complete this base two cross-number.

Across

1. 1011 + 11
101. 1110 ÷ 10
110. 1 + 1
111. 10 + 1

1000. 10000 ÷ 100
1010. 100 − 1
1011. 111 − 10
1101. 111 × 11

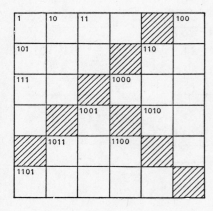

Down

1. 111 × 10
10. 1 + 10 + 100
11. 1·1 × 10
100. 1101 + 110
110. 11 + 10
1001. 1111 − 1010
1011. 0·1 + 0·1
1100. 100 ÷ 10

4 All these questions are in base twelve, give your answers also in base twelve.

(a) 29 + 37; (b) 46 + 6; (c) 25 + 27;
(d) T + T;   (e) T × 2;   (f) 12 × 11.

Base two fractions

Exercise H (Miscellaneous)

1 See Figure I.

2 See Figure I.

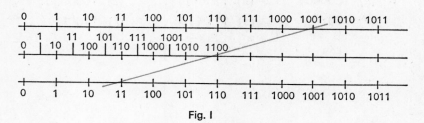

Fig. I

3

Fig. J

4 (a) 64; (b) 50; (c) 50;
  (d) 18; (e) 18; (f) 132.

Binary and duodecimal bases

6   (a) 16;   (b) 64.

7   (a) 1001;   (b) 1110;   (c) 10101;
    (d) 11010;   (e) 11111;   (f) 111111.

8   (a) 100000;   (b) 1100111;   (c) 10000001;
    (d) 110;   (e) 111·110;   (f) 11001.

Base two fractions

5 Design a set of addition scales which will work in base twelve.

6   8 Great-grandparents   4 Grandparents
       2 Parents   1 Person

Grandmother   *Grandfather*   *Grandmother*   *Grandfather*

         Mother   *Father*
                You

Here is part of your family tree; you should recognize the pattern of numbers by now.
    (*a*) How many great-great grandparents do you have?
    (*b*) How many great-great-great-great grandparents have you?

7 These numbers are written in base ten, convert them to base two.
    (*a*) 9;   (*b*) 14;   (*c*) 21;   (*d*) 26;   (*e*) 31;   (*f*) 63.

8 These questions are in binary; give your answers in binary.
    (*a*) 10101 + 1011;   (*b*) 1001011 + 11100;
    (*c*) 1010000 + 110001;   (*d*) 1101 − 111;
    (*e*) 1000·101 − 0·111;   (*f*) 101 × 101.

9. Statistics

1. A SURVEY

Let us imagine that a Headteacher wants to find out how his pupils come to school and how many use each method of travel.

Some will walk, some cycle, others come by train, and so on, and the Headteacher must first decide which headings to group pupils under. He might decide to try:

        Walk   Cycle   Bus   Train   Brought by Car

Already there are difficulties, and decisions will have to be made.

What about the pupil who has a long walk from the bus stop? Does he come under the heading 'Walk' or 'Bus', or should we make a new heading?

What about the girl who cycles when it is fine weather but who is brought by car when it is raining? You can probably think of other difficulties as well.

The Headteacher will probably decide to put these 'awkward' cases into the group they belong to most frequently, although in many cases this might not be an easy decision to make.

This means that the results will not give a perfectly accurate picture of the situation. Two people faced with the same decision might decide differently.

9. Statistics

We are constantly being bombarded with so-called statistics from newspapers, advertisement hoardings and television screens, and for this reason, if for no other, a fuller and more general treatment is needed today than hitherto.

It is very likely that many pupils will have met statistics before; some will have collected, tabulated or pictured data and drawn various sorts of chart in the junior school. It is also likely that most of this work will have been personally centred on them, with details of such things as heights, weights, favourite colours, etc., of themselves and their friends. It is because they can work with data that concern and interest them, that many pupils, who otherwise find mathematics difficult or distasteful, will settle down to statistical work.

Although a lot of time can be spent in collecting genuine data, the interest and mathematical disciplines involved make it well worthwhile. The collection and presentation of real data often involves group work and cooperative effort, and this aspect of working together is another point in favour of the inclusion of statistics.

All 'modern' syllabuses, both O-level and C.S.E., include the study of statistics, some even offering it as one of the main alternative subjects. At least one Board has a complete O-level in the subject. We believe that the trend to spend even more time on this subject should be encouraged.

1. A SURVEY

Statistics has been described as the collection, presentation, analysis and interpretation of numerical data. This first chapter dwells on the first two parts of this definition—collection and presentation.

As a background to this work, the class can be encouraged to collect and display actual charts taken from newspapers and magazines, but the teacher should be careful to select only the good ones at this stage and keep the doubtful or misleading ones for criticism at a later date.

It is suggested that the use of the word 'chart' be encouraged, and the word 'graph' reserved for line charts.

Statistics

2. BAR CHARTS

This is a very quick and easy way of presenting information. Do not use the word histogram. We shall use this word to refer to a diagram that shows frequency densities and which represents frequencies by areas and not heights of columns. Pupils should be reminded that a bar chart without a heading and a labelled frequency scale is meaningless.

A gap between bars is not essential, but it does make them stand out better, and the use of colour or colours can also have this effect. The order in which the bars are placed does not matter, unless one is comparing two bar charts in which case the relative order of the two should be the same. If the description of each bar is rather long, it is often clearer to write inside a horizontal bar than underneath a vertical one. Another reason for choosing to draw horizontal rather than vertical bars, or vice versa, could be that the scale worked out more conveniently in one direction than the other. (This scale should, above all, be easy to work from while allowing as much as possible for the fullest use of available space.)

A survey

Here is the result of a survey made on a class of first-form boys and girls to find out how many are using each method:

walk, 13; cycle, 8; bus, 4; train, 9; car, 2.

We can just write the results out like this, or we could try to find a way to display them so that they were easier to follow, or even to see at a glance.

The first method suggested is a simple table:

Walk	Cycle	Bus	Train	Car	
13	8	4	9	2	Total 36

The total has been included and this acts as a useful check.

2. BAR CHARTS

Another simple way to display this information is on a 'Bar Chart'.

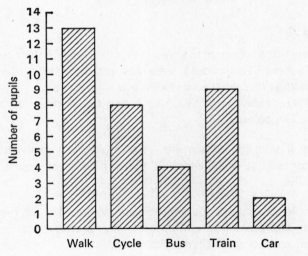

How 1a come to school

Notice: (i) The chart has a heading.
(ii) The words 'Number of Pupils' show what the numbers stand for.
(iii) The bars are not touching.

Statistics

A bar chart can be drawn sideways if you like:

How 1a come to school

| Car |
| Train |
| Bus |
| Cycle |
| Walk |

0 1 2 3 4 5 6 7 8 9 10 11 12 13 14
Number of pupils

In this case it is easier to write the words in the actual bars.

Exercise A

1 Carry out a survey of the way in which people in your own class travel to school. You should have a short discussion first to agree on headings and what to do with 'awkward' cases.
   Present your information both as a table and as a bar chart which you can colour.

2 See if you can obtain the same information as in Question 1 from another First Form which is the same size as yours, and compare results.

3 Would you expect any great difference between the results from your form and those from a fifth form? Try to obtain details from a senior form and compare them with yours.

4 Decide on headings to cover 'How my day is spent': for example, 'Working', 'Travelling', 'Eating' and so on. Try to avoid having more than 8 headings. Having decided upon headings which will be the same for all the class, each pupil should find out how long he spends on each activity and display his results in a bar chart.

Bar charts

The frequency axis need not start at zero, but when it does not, this should be made quite clear. It is the failure to do this (intentional or otherwise), that leads to many misrepresentations of data. However, although this is a possible discussion point, it should not be laboured as it will be raised again in a later book when pupils will be more experienced and better equipped to appreciate and discuss the issue.

Many of these technical details are best put across by question and discussion. For example:

1. Is it best to leave gaps between bars?
2. Does the order in which the bars are placed matter?
3. What are the merits of the sideways chart versus the upright one?
4. Should all the bars be the same colour, or can different colours be used?
5. Do we always have to start at zero?

Statistics

3. PIE CHARTS

Pie charts are much more difficult to produce than bar charts, but some pupils enjoy drawing them. They give good practice at long division and the use of the protractor. The examples in this text have been kept simple to avoid difficult arithmetic. The teacher should keep an eye on the arithmetic when pupils are asked to draw pie charts illustrating data of their own. Pupils should be encouraged to show clearly the associated working.

Most pupils derive pleasure and satisfaction from colouring the different sectors, which is really necessary if a pie chart is to have maximum impact. The order of the sectors is unimportant unless two pie charts are being compared, in which case the relative order between the two should be the same.

Bar charts

Problems arising:
(a) Shall we all do the same day?
(b) Should we choose a working day or holiday?

5 Make a list of all the subjects on your time-table and the number of minutes spent on each in the course of the week. Display these results in the form of a bar chart.

3. PIE CHARTS

Another simple way of displaying the relation between the different methods of travel and the numbers who use them, is by means of a 'Pie Chart'.

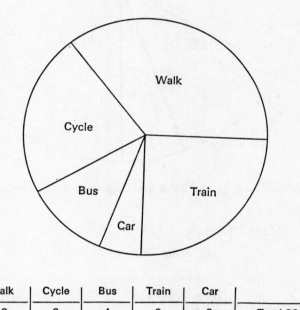

Walk	Cycle	Bus	Train	Car	
13	8	4	9	2	Total 36

Take the results of the survey on coming to school on p. 107. Each of the 36 pupils has to have an equal share of the pie. If 36 pupils share 360° then each one should have 10°. In degrees, the table will be:

Walk	Cycle	Bus	Train	Car	
130	80	40	90	20	Total 360

Use your protractor to check the above diagram.

T 241

Statistics

If a large pie chart is made then everybody's name can be included:

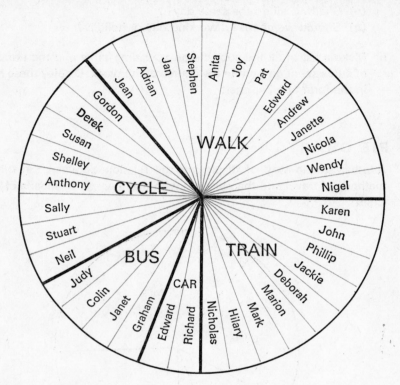

Colours will help to make the five important sectors stand out.

4. PICTOGRAMS

How 1a come to school

4. PICTOGRAMS

As presented here, a pictogram is really an eye-catching form of bar chart, and has more popular appeal than the latter. Its use is straightforward when the population is small, when each motif can represent one member. As soon as this becomes impracticable, obvious difficulties arise, with the necessity of drawing a given fraction of the motif. (It is worth discussing this with any classes capable of appreciating this difficulty and disadvantage of the pictogram.) Although there are no questions requiring a pictogram in Exercise B, by this stage, pupils should have collected plenty of data suitable for this means of illustration.

The motifs should be simple, of the same size, and, where possible, suitably related to the data being illustrated. They should be equally spaced apart and aligned vertically. Strictly speaking, it should be made clear what one motif represents,

(e.g. 👤 = 1 vote).

Exercise B

4   Sectors are:

Rent and Rates	112°
Heating and Lighting	48°
Food and Drink	104°
Clothing	20°
Travel	32°
Entertainment	44°

5. PROJECTS

One of the great advantages of statistics as a practical subject is that it is usually possible to understand the purpose behind the collection and the presentation of data. In the project work, the purpose should be discussed because it not only provides a motive for the work, but also informs and directs the project itself.

In this topic, perhaps more than any other, there is need for considerable preparation and preliminary discussion before setting out on a task. It should be accepted that there will be times when the first attempt will have to be scrapped and the whole experiment carried out again in the light of the knowledge gained. As a guide to the sort of preparation that is necessary, the first two projects have been described in some detail; it is hoped that they will provide a pattern for the subsequent projects.

Project 2

The order of frequency of letters in the English language is

E T O A N I R S H D L C W U M F Y G P B V K X Q J Z.

With only a small sample passage there can be considerable variation.

Pictograms

An attractive way of representing information is the *pictogram*. In this, little drawings are used to display the details.

A stencil, lino-cut or potato-cut can be used to help produce these diagrams.

Compare this chart with the sideways bar chart on p. 108.

Exercise B

1. Make a pie chart of the results of the survey on 'Travelling to school' in your class. Remember that 360° have to be shared out equally by the total number of pupils in the survey.
2. Make a pie chart to show 'How my day is spent'.
3. Show your results to Exercise A, Question 5, as a pie chart.
4. During one month, this is the way a man spent his money:

	£
Rent	24
Rates	4
Heating	8
Lighting	4
Food	20
Drink	6
Clothing	5
Travel	8
Entertainment	11

If all these items were represented separately on a pie chart it would look too complicated and lose its effect. Group together (Rent and Rates), (Heating and Lighting), and (Food and Drink), and then make a pie chart.

5. PROJECTS

1. A vehicle survey

What could you do to convince people that there ought to be a pedestrian crossing and a crossing patrol warden outside your school?

If a case is to be presented to the authorities it would be better if it was backed by details of the sort of vehicles that passed and their numbers, so a proper traffic survey is called for.

Many things have first to be decided and much preliminary discussion will be needed.

Points for discussion and decision are:

(*a*) Which day of the week will you choose? Remember early closing.
(*b*) What time of day will you choose?

Statistics

(c) For what length of time will you count vehicles?
(d) Will you repeat the experiment over many days, say a week?
(e) Will you just do the same day each week for four weeks?
(f) Where will you stand?
(g) Will you count traffic in one direction only or in both?
(h) How will you classify the types of vehicle?
(i) How will the weather affect things?
(j) Is it necessary to produce a form for recording results?
(k) Is a trial run desirable?

Having decided on all these points, and any others which might arise because of the special circumstances of your school, you are ready to do the actual survey. Here is a suggested form:

VEHICLE SURVEY

Name .... Date ....

Location .... Time started ....

Direction .... Time finished ....

Weather conditions ....

	TALLY	TOTAL
Motor cars and vans		
Lorries and large commercial vehicles		
Buses and coaches		
Motor cycles, mopeds, scooters		
Pedal cycles		
Other vehicles (with description)		

Projects

Others could be—Fire engines, tractors, police and military vehicles, horses and carts, ambulances, steam rollers, etc.

The attempt to classify cars and vans into one group and lorries and large commercial vehicles into another might lead to some difficulties which you will have to try to overcome.

Results:

(i) A neat and completed form can serve as a table.
(ii) Produce a bar chart.

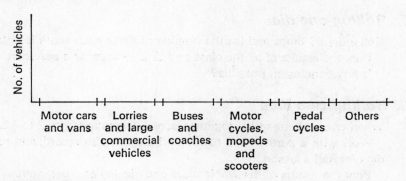

(iii) Display as a pie chart.
(iv) Display as a pictogram.

2. The most common letter

Aim: To list the order of frequency of letters in an English book.

Method:

(a) Decide on a book which every member of the class has.
(b) Assign a page to each pupil.
(c) List the alphabet and tally off through the page.
(d) Pool results.

Results:

(a) Display as a table from A to Z.
(b) Re-arrange the table and list the letters in order of frequency.
(c) Display as a bar chart in order of frequency.

Why is a pie chart unsuitable?

Conclusions:

What is the most common letter you have discovered?
Would this have been the same if you had chosen any other book?
Can these facts be put to use by anybody?

Statistics

3. The most common letter in French
Repeat Project 2 but use a French book.

4. The most common length of word
Carry out a survey to discover the most common length of word in a book, following the lines of Project 2.
How do you think the result depends upon the type of book chosen?

5. Rolling one die
Roll a die 60 times and list the number of times each score is obtained.
Pool the results of all the class and display them as a bar chart.
Is any conclusion possible?

6. Rolling two dice
When two dice are rolled together, scores can range from 2 to 12.
Work with a partner, one to roll and the other to record, and roll two dice for half a lesson.
Pool the results of the whole class and display as a bar chart.
What about a conclusion?

Miscellaneous Exercise

1 Describe the data shown in this bar chart. How many families are involved? How many children?
   Draw a similar chart for the families represented in your class. Discuss whether the same shape would be expected for a similar class 100 years ago.

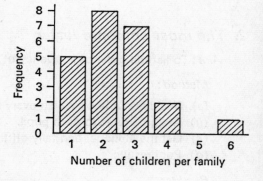

2 A shopkeeper sold fireworks at various prices. There were 1p ones, 2p ones, etc. He kept a record of the first 60 fireworks bought one day as follows:

1p	1p	6p	3p	2p	3p	6p	9p	2p	4p	2p	4p
6p	2p	1p	4p	9p	5p	4p	1p	2p	2p	1p	4p
5p	3p	3p	3p	9p	6p	1p	3p	6p	4p	1p	5p
2p	4p	6p	2p	3p	2p	1p	1p	9p	6p	2p	1p
1p	2p	6p	1p	5p	3p	1p	1p	9p	1p	2p	1p

114

Projects

Project 3

The order of frequency of letters in the French language is

E N A S R I U T O L D C M P V F B G X H Q Y Z J K W.

Project 5

A large number of results should give roughly equal frequencies for each score. This is an example of a rectangular distribution.

Project 6

A score of 7 should have the greatest frequency.

This is a triangular distribution, the frequencies going up by equal steps to the frequency of score 7 and then down again.

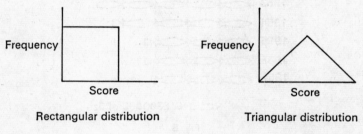

Fig. A

Miscellaneous Exercise

1   23 families.   56 children.

2   16 @ 1p, 12 @ 2p, 8 @ 3p, 7 @ 4p, 4 @ 5p, 8 @ 6p, 5 @ 9p.

Statistics

3 (*b*) Sectors are:

Earwigs	144°
Caterwaulers	108°
Crotchets	48°
Popalongs	36°
Squareboys	24°

4 L, G, O and Y are the four most frequently occurring letters. This is a Welsh place name.

5 Sectors are:

Cattle	90°
Sheep	225°
Pigs	45°

6 See Figure B.

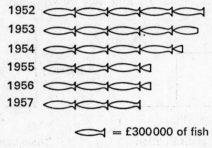

Fig. B

Bibliography

Starting Statistics, Lewis and Ward, Longmans.

This book contains many practical projects and experiments for those who desire a fuller treatment.

Projects

(a) Put these results into a table, using tally marks to show how many of each type were sold.

(b) Display these results as a bar chart.

(c) Which firework was the most popular?

3   A musical survey in form 2Z revealed the following facts about the popularity of certain recording groups.

   12 liked the Earwigs best of all,   3 liked the Popalongs,
   9 liked the Caterwaulers,   2 liked the Squareboys.
   4 liked the Crotchet String Quartet,

   Illustrate this information by drawing:

   (a) a bar chart;   (b) a pie chart;
   (c) devise some other method of showing the information.

4   Make a frequency table for the letters occurring in the following place-name:

   LLANFAIRPWLLGWYNGYLLGOGERYCHWYRNDROBWLL-
   LLANTYSILIOGOGOGOCH

   Which are the four most frequently appearing letters? In which country do you think this place is?

5   The numbers of principal farm animals in Great Britain are:

   cattle, 12 million;   sheep, 30 million;   pigs, 6 million.

   Draw a pie chart to illustrate this information and state the angle of each 'slice' in degrees.

6   Draw a suitable pictogram to illustrate the following figures, giving the value in hundreds of thousands of £'s of herring landed in English and Welsh ports in the years shown:

1952	...	15	1955	...	10
1953	...	14½	1956	...	10
1954	...	13	1957	...	9

7   Copy this table and see how many points you can list:

Bar chart		Pie chart	
Advantages	Disadvantages	Advantages	Disadvantages

Interlude

'SHIFTS'

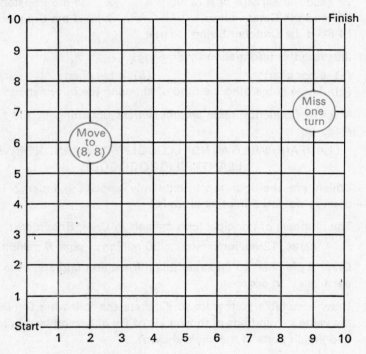

For the game of 'Shifts' you need: squared paper marked as shown, two small cubes and some counters. The cubes should be marked

$$X0, \quad X1, \quad X2, \quad X3, \quad X4, \quad X5$$

and

$$Y0, \quad Y1, \quad Y2, \quad Y3, \quad Y4, \quad Y5.$$

Two or more people can play, each one taking turns at throwing the two marked cubes and at moving their counters. The numbers marked on the cubes state the distance to be moved or 'shifted' towards the right and towards the top of the board. A player at (2, 1) who throws (X1, Y4), moves to (3, 5). A player at (8, 9) with this throw would move to (9, 9) ignoring the Y shift that takes him beyond the edge of the board.

Four or five more instruction boxes should be added before starting the game. The object of the game is to move from the start (0, 0) to the finish (10, 10) before the other players.

T252

Interlude

SHIFTS

In the next two books of the course, we shall be considering translations as movements about a plane or space, and vectors, i.e. ordered pairs or triples, to describe the translations. This Interlude is intended to provide a foundation for this idea. The board will provide an approximation to a coordinate system and the readings of the dice will provide statements of movements across and up the board.

At this stage, no negative numbers are used.

10. Directed numbers

We shall attempt to show the great convenience that is to be found in some situations when the system containing only positive numbers is replaced by another system containing both positive and negative, i.e. directed numbers.

In *Book C*, there will be a chapter on the combination of directed numbers under addition and subtraction. Difficulties are always liable to arise when we wish to establish the results to be expected when directed numbers are combined. We hope that, in this chapter, we can lessen this liability in three ways:

(i) by emphasizing the use of the number line in displaying directed numbers;

(ii) by introducing the 'upper height' notation to distinguish between, for example, 'negative 2' written as '$^-2$', and 'from k subtract 2' which would be written as '$k-2$';

(The upper height negative notation will be used throughout the course, but the upper height positive notation only until *Book D*.

(iii) by making it clear that we need *differences* between numbers in the 'positive number' system in order to obtain the numbers of the 'directed' system.

In this chapter, we have immediately 'tied' the new 'difference' numbers to an origin and so formed a new number line extending in both directions from the origin. We shall later consider the new 'difference' numbers as shifts over the number line; in this way we can more clearly interpret, for example, subtraction or multiplication of directed numbers.

1. REFERENCE POINTS

This section brings out the idea of an arbitrary zero or reference point, the need to be able to differentiate between points on one side of the reference point and the other, and the need to subtract in order to obtain the new table of intervals. No mention is yet made of a 'negative number'. Figures are not drawn to scale as, for the moment, it is relative order along a line that we are concerned with. The number line is introduced in Section 3.

10. Directed numbers

1. REFERENCE POINTS
1.1 Time-tables

Here is a time-table showing part of a schoolboy's morning.

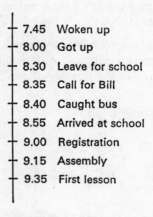

7.45	Woken up
8.00	Got up
8.30	Leave for school
8.35	Call for Bill
8.40	Caught bus
8.55	Arrived at school
9.00	Registration
9.15	Assembly
9.35	First lesson

Fig. 1

As the boy finds it difficult to get up and get to school on time, he has a time-table pinned up above his bed. In order to help him get there on time, he wants to know how many minutes he has left before 'zero hour', that is, Registration Time. His time-table therefore looks like this:

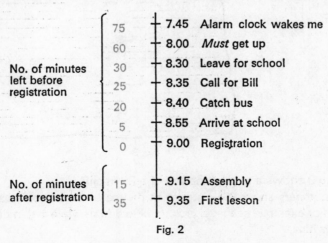

Fig. 2

Directed numbers

Exercise A

1 (a) Copy Figure 1 into your book and leave space to put the following events in their correct place:
   (i) 8.15: breakfast;
   (ii) 8.25: clean shoes;
   (iii) 8.50: get off bus;
   (iv) 9.10: go down to Assembly;
   (v) 9.30: Assembly finishes.

   (b) Taking arrival time at school as 'zero hour', show the number of minutes left before arrival or the number of minutes after arrival, for each event.

2 Draw up a time-table as in Figure 1 showing what you do yourself in the mornings. Decide what will be your 'zero hour' and for each event put in the number of minutes before or after this zero hour.

1.2 Distances

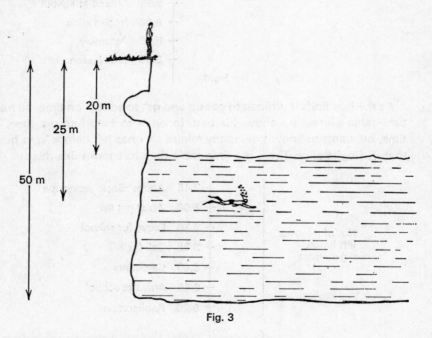

Fig. 3

Figure 3 shows a man at the top of a 20 m cliff.
The numbers show the distance in metres of certain objects below his feet. For example, sea level is 20 m below him, while a skin diver is 25 m below him.

Reference points

1.1 Time-tables
Exercise A
1

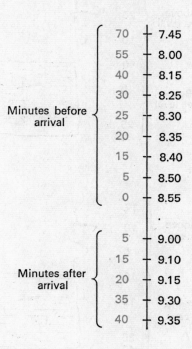

Directed numbers

1.2 Distances

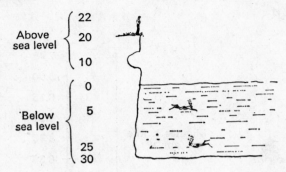

Above sea level { 22, 20, 10
Below sea level { 0, 5, 25, 30

Exercise B

1

Above cave { 12, 10, 0
Below cave { 10, 15, 35, 40

2   Scafell Pike   60 m below C
    Snowdon   60 m above C
    Ben Nevis   300 m above C
    Slieve Donard   190 m below C

3   Nottingham   1° S.
    Newcastle   1° N.
    Edinburgh   2° N.
    Londonderry   1° S.
    Southampton   3° S.

Reference points

Copy Figure 3 but instead of making the top of the cliff your 'zero' or 'reference point' make sea level your zero and put in the number of metres above or below it of the top of the cliff, the skin diver and the sea bed. Mark in also
(a) a cave, 10 m below the cliff top;
(b) the head of the man, 2 m above the cliff top;
(c) another diver, swimming 5 m above the sea bed.

1.3 Class project on ages

Find out the ages to the nearest month of each person in your class. Taking your own age as your 'reference point' or 'zero', write down the ages of everybody else as so many months older than or younger than yourself.

You may find it easier if you first rewrite your list of names in order of age, with the youngest first and oldest last.

Exercise B

1 Copy Figure 3 again but this time make the entrance to the cave your zero reference point.

2

Mountains	Height above sea level (m)	
Scafell Pike	980	
Snowdon	1100	
Ben Nevis	1340	
Slieve Donard	850	
Carrauntoohil	1040	0

Copy this table which shows the heights of the highest mountain in each country of the British Isles.

(a) Taking Carrauntoohil as your zero, calculate in the heights above and below this reference point.
(b) Choose any other mountain you like as your zero. Add another column to your table and fill in the heights of the other mountains above and below this new reference point.

3 If we know how far north or south of the Equator a place is, we can begin to describe its position. In this case, we use the Equator as our zero. For example, Nottingham is 53° N, York is 54° N, Newcastle is 55° N, Edinburgh is 56° N, Londonderry is 53° N, Southampton is 51° N.

Directed numbers

Taking York as your zero, write down the position of these places north and south of this new zero.

4   In describing the positions of various places, a line through the Poles and Greenwich is taken as zero. A place is so many degrees west or east of this line. For example, Liverpool is 3° W, Birmingham is 2° W, Swansea is 4° W, Portsmouth is 1° W, Canterbury is 1° E, Glasgow is 4° W, and Belfast is 6° W.

Taking a line through Birmingham as your zero, write down the position of the other places, including Greenwich, west and east of this new zero.

5   Figure 4 shows the average body temperatures of some species of animal, measured on the Celsius scale. Copy the table and show the number of degrees Celsius that each of these temperatures is above or below the body temperature of man.

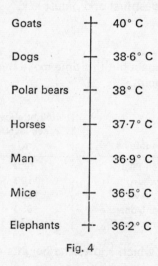

Fig. 4

2. DIRECTED NUMBERS

In the examples of the previous section, all measurements were made from a reference point, sometimes called the 'zero'. Any point can be chosen as a zero; it is a matter of convenience.

We have used words such as 'before' and 'after', 'above' and 'below', 'east' and 'west', 'north' and 'south', 'younger' and 'older'. These words tell us the direction of other readings on either side of the zero.

This pattern appears again and again so a new set of numbers has been invented. This set can be used in all these situations and is called the set of 'directed numbers'. Numbers on one side of the zero will be called

Reference points

4   Liverpool   1° W.
    Swansea   2° W.
    Portsmouth   1° E.
    Canterbury   3° E.
    Glasgow   2° W.
    Belfast   4° W.
    Greenwich   2° E.

5

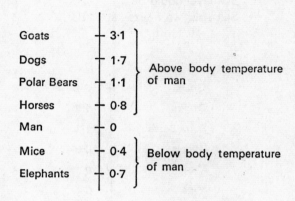

2. DIRECTED NUMBERS

So far, the majority of pupils will have always taken the smaller number from the larger and then described the result as so much above or below the reference point.

This process continues with the introduction of the names positive and negative so that, for example, the sequence 13, 12, 11 with a new reference point at 12 will be argued as:

13   gives $13 - 12 = 1$ and this will be called $^{+}1$;
12   is the new reference point and will be called 0;
11   gives $12 - 11 = 1$ and this, on the opposite side, will be called $^{-}1$.

Directed numbers

We feel that, as soon as this is understood, the more general method should be introduced in which the new numbers are obtained directly by always subtracting the new reference point from the other numbers whether they are larger or smaller. This means that the second and third lines would read simply

$$12 - 12 = 0;$$
$$11 - 12 = {}^-1.$$

Sea level could be $^+20$.
Salt solution freezes at $^-23$.

Exercise C

1

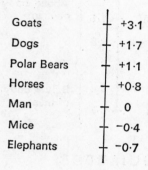

Goats	+3·1
Dogs	+1·7
Polar Bears	+1·1
Horses	+0·8
Man	0
Mice	−0·4
Elephants	−0·7

2   (a) $^+700$;   (b) $^+120$;   (c) $^-100$;
    (d) $^-15$;   (e) $^-140$.

Directed numbers

positive and will carry a '+'; numbers on the other side will be called negative and will carry a '−'.

For example, on Figure 2 with the zero at 9.00, arrival at school can be taken as −5 and Assembly as +15. In Figure 3, taking the top of the cliff as zero, the man's head is at +2 and the sea level at −20.

Sometimes it does not matter which direction is taken as positive and which as negative. If the man's head is taken as −2, what would the sea level be? But on some occasions, it is customary to accept one direction as positive. All the temperatures on Figure 4 are positive. How would you give the temperature at which a salt solution freezes if this is 23° below zero?

2.1 Class project on heights

Find out the heights to the nearest centimetre of each person in your class. Taking your own height as your zero, write down the heights of everybody else in relation to your own. Use positive numbers to show the number of centimetres taller than yourself and negative numbers to show the number of centimetres shorter than yourself. You may find it easier to rewrite your list of names in order of height with the smallest first and the tallest last. Make a plan of your final result using a line like the one in Figure 5.

Exercise C

1 Copy Figure 4. Take the average body temperature of man as your zero and mark the other points on the scale using positive numbers for points above your new zero, and negative numbers for points below.

2 The height of John's house is 100 m above sea level. Describe the heights of the following objects in relation to John's house:
   (a) a television mast 800 m above sea level;
   (b) a church   220 m above sea level;
   (c) a boat at sea   0 m;
   (d) a bridge   85 m above sea level;
   (e) a submarine   40 m below sea level.

Directed numbers

3 (a) If 9 o'clock is zero hour for your arrival at school in the mornings, express the following arrival times in minutes making use of the extended number system. (For example, a quarter to nine is $^-15$ min.)

(i) 3 min to 9 o'clock; (ii) 5 min past 9 o'clock; (iii) 8.50 a.m.

(b) What would your answers be if the zero hour were 8.55 a.m.?

4 The average age of a class is 12 years 3 months.

(a) Judith, John, Jack and Jim are 12 years, 11 years 10 months, 12 years 7 months, and 11 years 8 months old respectively. How do their ages differ from the average?

(b) Barbara, Betty, Benjamin and Bill have ages which, when related to the average, are $^-9$ months, 0 months, $^+3$ months, and $^-5$ months. What are their ages?

(c) Name two members of the class with the same age.

3. THE NUMBER LINE

We have already met the number line as in Figure 5 made up of the counting numbers and zero.

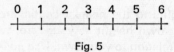

Fig. 5

Suppose we now wish to extend the number line to the left of zero. This will give us numbers on both sizes of zero. In order to be able to tell the difference between them, we can use the positive and negative signs. The numbers to the right of zero will carry the positive sign and the numbers to the left of zero will carry the negative sign, as in Figure 6.

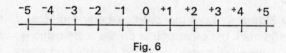

Fig. 6

If you have done the class project on heights properly, you will have already shown the information you have gathered on a number line. Check to see whether it is correct.

Now show the information gathered on class ages on a number line, with your own age as your zero.

On your number line for class heights, one person is represented by $^+3$, and another by $^-2$. Which is the taller?

Directed numbers

3 (*a*) (i) ⁻3, (ii) ⁺5, (iii) ⁻10;
   (*b*) (i) ⁺2, (ii) ⁺10, (iii) ⁻5.

4 (*a*) Judith   ⁻3 months   (*b*) Barbara   11 years 6 months
       John   ⁻5 months   Betty   12 years 3 months
       Jack   ⁺4 months   Benjamin 12 years 6 months
       Jim   ⁻7 months   Bill   11 years 10 months
   (*c*) John and Bill.

3. THE NUMBER LINE

The taller person is represented by ⁺3.

Directed numbers

The older person is represented by $^+1$.
$^-20°$ is hotter than $^-30°$.
$^-5$ is smaller than $^+3$.

$$^-2 \not< {}^-4, \quad {}^-4 < {}^-1, \quad {}^-1 < {}^+3.$$

(a) $^-4, {}^-5, {}^-6, \ldots$;
(b) $^-5, {}^-4, {}^-3, \ldots, {}^+3, {}^+4$;
(c) $^+1\frac{1}{2}, {}^+1\frac{1}{4}, {}^+1\frac{3}{4}, \ldots$.

Exercise D

1. (a) $^+3 < {}^+7$;   (b) $0 < {}^+4$;   (c) $^-2 < {}^+4$;
   (d) $^-3 < 0$;   (e) $^-3 < {}^-1$;   (f) $^-5 < {}^-3$.

2. $^-10, {}^-8, {}^-5, {}^-3, {}^+2, {}^+6, {}^+7$.

3. (a) $^+4 < {}^+6$;   (b) $^+5 > {}^+3$;   (c) $0 < {}^+6$;
   (d) $^-3 < 0$;   (e) $^-4 < {}^+2$;   (f) $^+5 > {}^-3$;
   (g) $^-2 > {}^-3$;   (h) $^-5 < {}^-1$;   (i) $^-1 < {}^+1$.

4. There are many possible answers here.

The number line

On your number line for class ages, one person is represented by $^+1$ and another by $^-4$. Which is the older?

In the Celsius scale of temperature, which is hotter, $^-20°$ or $^-30°$?

Which is smaller, $^-5$ or $^+3$? (see Figure 6).

The answers to all these questions depend on the *order* of the numbers. On the number line, shown in Figure 6, $^-5$ is smaller than $^+3$ because $^-5$ is to the left of $^+3$. You have already met a symbol showing this order relation 'is smaller than', namely '<'. So we write

$$^-5 < {^+3}.$$

Mark the numbers on a number line and find out which of the following statements are true.

(a) $^-2 < {^-4}$;   (b) $^-4 < {^-1}$;   (c) $^-1 < {^+3}$.

We can write these last two relations together as

$$^-4 < {^-1} < {^+3}$$

because $^-1$ is to the left of $^+3$ and to the right of $^-4$.

Give some possible values for x if:

(a) $x < {^-3}$;   (b) $^-6 < x < {^+5}$;   (c) $^+1 < x < {^+2}$.

Exercise D

1 Write each pair of numbers with the smaller first, for example, $^+2 < {^+5}$:

(a) $^+3, {^+7}$;   (b) $^+4, 0$;   (c) $^+4, {^-2}$;
(d) $0, {^-3}$;   (e) $^-1, {^-3}$;   (f) $^-5, {^-3}$.

2 Arrange the following numbers in order, smallest first:

$$^-3, {^+2}, {^-5}, {^+7}, {^+6}, {^-8}, {^-10}.$$

3 Copy each pair of numbers and put in the correct sign, (either < or >) between them:

(a) $^+4, {^+6}$;   (b) $^+5, {^+3}$;   (c) $0, {^+6}$;
(d) $^-3, 0$;   (e) $^-4, {^+2}$;   (f) $^+5, {^-3}$;
(g) $^-2, {^-3}$;   (h) $^-5, {^-1}$;   (i) $^-1, {^+1}$.

4 Give two possible values of x if:

(a) $x < {^+10}$;   (b) $^+7 < x$;   (c) $x < {^-4}$;
(d) $^-8 < x$;   (e) $x > {^+2}$;   (f) $^-3 > x$.

T 267

Directed numbers

5   Part of the set of points for which $x < {}^+5$ can be shown on a number line in the following way:

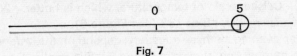

Fig. 7

The circle shows that the point $^+5$ is not included in the set. In a similar way, show on the number line the sets of points for which:

(a) $x < {}^+3$;   (b) $x < {}^-1$;   (c) $^+2 < x$;
(d) $^-3 < x$;   (e) $x > {}^-5$;   (f) $^+1 > x$.

6   Give a value for x if:

(a) $^+2 < x < {}^+9$;   (b) $^-3 < x < 0$;   (c) $^-7 < x < {}^-4$;
(d) $^-2 < x < {}^-1$;   (e) $^+5 > x > {}^+1$;   (f) $^-3 > x > {}^-6$.

5

(*a*)
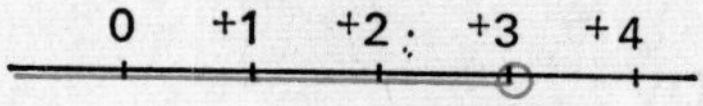

(*b*)
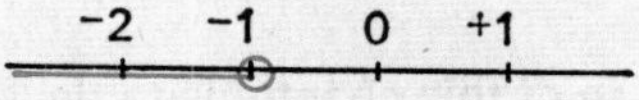

(*c*)
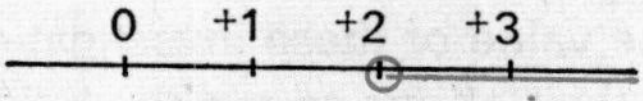

(*d*)
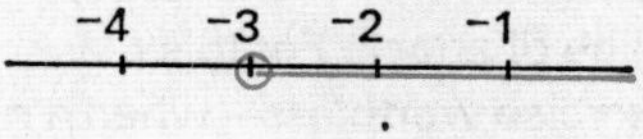

(*e*)
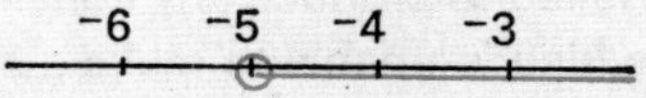

(*f*)
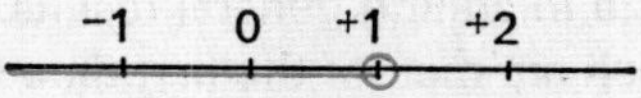

6 Pupils' most likely answers are:

(*a*) $^+3, ^+4, \ldots, ^+8$; (*b*) $^-2, ^-1$; (*c*) $^-6, ^-5$;

(*d*) $^-1\frac{1}{2}, ^-1\frac{3}{4}, \ldots$; (*e*) $^+4, ^+3, ^+2$; (*f*) $^-4, ^-5$.

11. Topology

The main purpose of this chapter is to develop geometrical intuition. The first five of the investigations which appear at intervals throughout the chapter are an essential part of the course and are particularly suitable for group work. The value of these investigations lies in the thought which pupils put into their efforts to seek solutions rather than in any results which may be obtained. Hints are rarely given in the text since these tend to prevent a free and inventive approach in that they restrict a pupil's thinking to one line of attack. For this reason we must not be disappointed —or think the exercise worthless—when a group of pupils fails to produce a complete and polished set of conclusions or, in the first instance, appears to have failed to produce any conclusions at all! A suggestion for aiding a group which is unable to make a start appears at the end of the comments on each investigation, but there is, of course, no one right suggestion. Several of the investigations offer opportunities for discussing the use of algebra in stating general results.

The ideas which are met for the first time in this chapter and which will occur again in later work are collected together in the summary on p. T303. These results are a record of what the pupil has done and no attempt need be made to learn them.

1. TOPOLOGICAL TRANSFORMATIONS

(a) and (b). These questions are intended for class discussion, but the discussion is sometimes more fruitful if it is preceded by short group discussions or individual consideration of the questions. There follows an example of such a discussion on (b):

- Q. In what way are the figures alike?
- A. They all have four spaces.
- Q. Spaces?
- A_1 They divide the page into four parts.
- A_2 Five parts.
- Q. Why five?
- A. There's the outside too.
- Q. Do all the drawings have five regions?
- A_1 Yes.

11. Topology

1. TOPOLOGICAL TRANSFORMATIONS

Look carefully at the drawings in Figure 1.

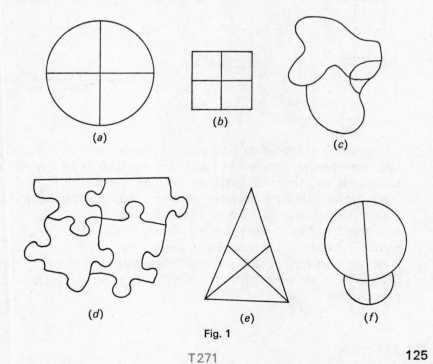

Fig. 1

Topology

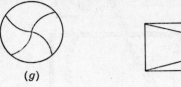 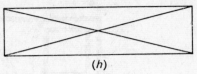

Fig. 1

(a) Make a list of the ways in which the drawings differ from one another. *Example*: some drawings are made entirely from segments of straight lines; others are not.

(b) Make a list of the ways in which the drawings are alike. *Example*: each drawing divides the page into the same number of cells or regions.

Imagine Figure 1 (b) to be drawn on a very thin rubber sheet. This can be pulled and twisted about as much as you like so long as you do not tear the rubber or stick two pieces together.

Figure 2 shows Figure 1 (b) drawn on a rectangular piece of rubber sheet which is pulled about so that the drawing looks like Figure 1 (a).

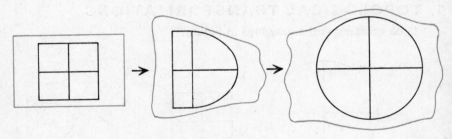

Fig. 2

Topology is about points, lines and the figures they make; but length, area, curvature and angle can be altered as much as you wish. Thus topology is sometimes called rubber-sheet geometry.

(c) Figure 3 shows a piece of wire netting. Suppose it to be twisted—even stretched—but not torn or fixed together. After this it will be a different shape; it may not be flat. But some things will remain unchanged. What are these? What differences will there be? Discuss your answers with your neighbour or your teacher.

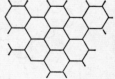

Fig. 3

Topological transformations

A₂ They are all joined up.
Q. What do you mean by joined up?

Pause

Q. Draw some figures that aren't joined up.
A₁

A₂

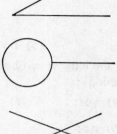

A₃

A₄ The ones in the book have a boundary; these haven't.
A₅ They have a boundary and two lines cross inside.
Q. Will any of these do?

A. No.
Q. But two lines cross inside!
A. The lines must join the boundary at different points.
Q. Draw some other figures like the ones in the book.

Pupils will probably express themselves in informal and imprecise language, but the discussion should bring out the following ideas:
  (1) distance, angle and direction are not preserved;
  (2) any one of the figures could be stretched and pulled into any one of the others if distance, angle and direction are ignored;
  (3) the number of lines meeting at a point and the order of points on a line do not change.

Topology

(c) This can be demonstrated practically. The sort of statements that remain true are: 'There are 31 joins from which 3 wires extend', 'There are 9 closed loops', 'There are 15 loose ends', etc. There will be changes in the length and breadth of the piece of netting, the angles between the wires and (theoretically) the total length of wire used—although stretching such wire is not a practical proposition.

Exercise A

1 (a) Yes;   (b) no;

   (c) no (the leg of the letter P can be shrunk to a very small size, but not reduced to nothing);

   (d) yes;   (e) yes;   (f) no;   (g) yes.

2 (a) Yes;   (b) no;

Topological transformations

(i) We call bending and stretching (but not tearing or sticking) *topological transformations*.

(ii) Facts that are still true about a drawing or network after it has been transformed are called *invariant* because they do not change or vary. The order of points on a line is invariant. The distance between two points is not invariant.

(iii) Two curves such that each is a topological transformation of the other are said to be *equivalent*.

Exercise A

1 A topological transformation of a circle is called a *simple closed curve*. Which of the drawings in Figure 4 are simple closed curves? (*Reminder*: you may not cut or stick.)

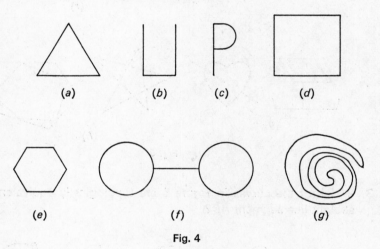

Fig. 4

2 Can the first curve of each pair in Figure 5 be transformed topologically into the second curve?

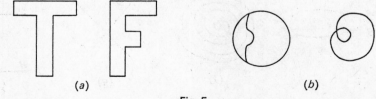

Fig. 5

T275

Topology

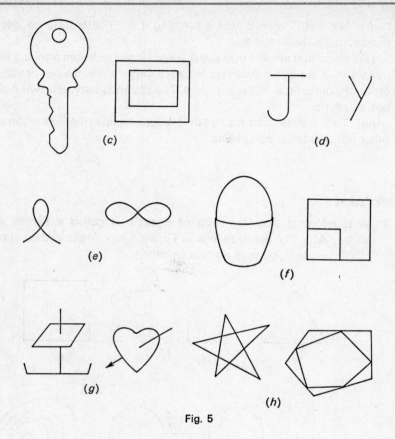

Fig. 5

3 Which of the curves in Figure 6 are topologically equivalent to the straight line segment AB?

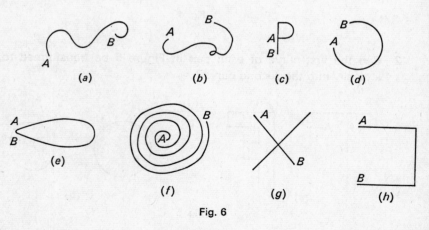

Fig. 6

(c) yes;	(d) yes;		
(e) no;	(f) yes;		
(g) yes;	(h) yes.		

An intuitive feeling that the first curve can be pulled and stretched into the second curve is sufficient at this stage.

3 (a) Yes;   (b) no;   (c) no;   (d) yes;
   (e) no;   (f) yes; (g) no;   (h) yes.

4 (a), (b) and (d) are equivalent, so are (c) and (e).

5 See Figure A.

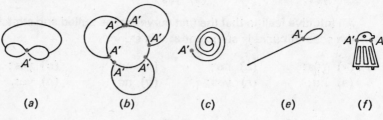

Fig. A

(a) A' is unique.
(b) A' can be in any of the six positions shown.
(c) A' is unique.
(d) The figures are not equivalent.

Topological transformations

4 Which of the drawings in Figure 7 are topologically equivalent to each other?

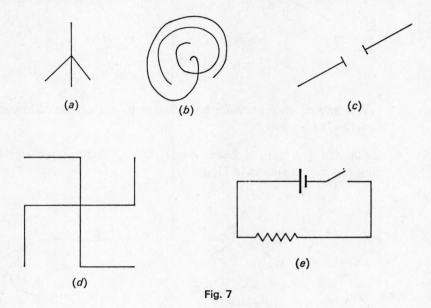

Fig. 7

5 Which of the pairs of drawings in Figure 8 are equivalent? If you think a pair is equivalent, copy the second member of the pair and mark on it a possible position for A', the image of A under the transformation. If more than one position is possible, mark all of them. You are allowed to turn the figures over.

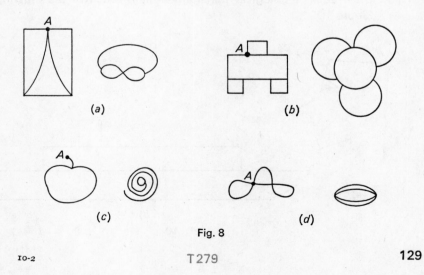

Fig. 8

Topology

Fig. 8

Will any of your answers be different if you are not allowed to turn the figures over?

6 What can you turn a beetle into? The original insect and two suggestions are shown in Figure 9.

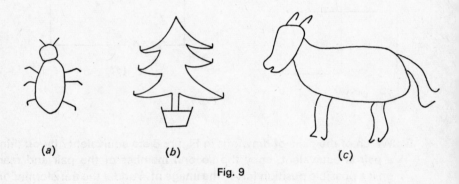

Fig. 9

7 Draw some topological transformations of each of the drawings in Figure 10.

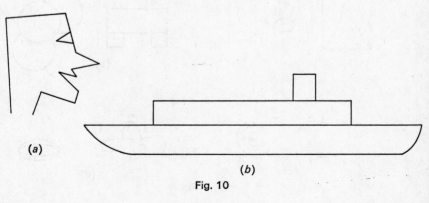

Fig. 10

Topological transformations

(e) A' is unique.
(f) A' can be in either of the two positions shown.
If turning the figures over is not allowed, then the number of answers to (b) and (f) is halved.

7 Figure B shows some pupils' suggestions for topological transformations of the drawing in Figure 10(b).

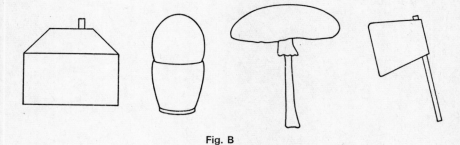

Fig. B

2. NODES

(*a*) One path leads from A to C; the other two lead from A to itself. Any circuit may be traversed in either direction, forming two distinct paths.

(*b*) One.

(*c*) C is a 1-node.

(*d*) Q, R, S, W are 1-nodes, P, V are 3-nodes; T is a 4-node, U is a 6-node.

Nodes

(c)   (d)

Fig. 10

2. NODES

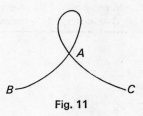

Fig. 11

(a) Look at Figure 11. There are four paths from the point A. One leads to B; where do the others lead?

(b) How many paths are there from C?

A point with at least one path leading from it is called a *node*. The *order* of the node is the number of paths. A is a 4-node because there are four paths from A.

(c) Describe the node at C.

(d) State the order of each node marked with a letter in Figure 12.

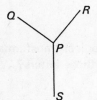

Fig. 12

T 283

Topology

Exercise B

1. Make a list of the letters in Figure 13 which have, amongst others:
   (a) one 3-node;   (b) two 3-nodes;   (c) one 4-node.

a b c d e f g h i j k l m n o p q r s t u v w x y z

Fig. 13

2. Why is it impossible to draw a figure with one 1-node and no other nodes?

3. Draw a line segment. Mark a 2-node on it. Mark another one. How many are there altogether?

4. Copy and complete the following table for the networks in Figure 14.

Figure	Total number of nodes	Number of 1-nodes	Number of 3-nodes	Number of 4-nodes	Number of 5-nodes
(a)					
(b)					
...					

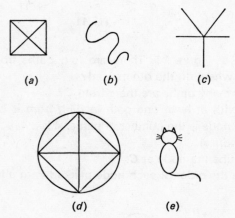

Fig. 14

5. Which of the letters in Figure 13 are topological transformations of the letter *C*? How many nodes has each of these letters? What kind of nodes are they?

6. Which of the letters in Figure 13 are equivalent to the letter *y*?

Nodes

Exercise B

With more able pupils this exercise could be omitted and the important ideas of Questions 2 and 3 brought out in dealing with Investigation 1.

1. (a) *a, b, d, e, g, h, n, p, q, r, t, y* (as printed in the text) have exactly one 3-node;
   (b) *m* has exactly two 3-nodes;
   (c) *f, k, x* have one 4-node.

2. The path from the 1-node must lead to another node.

3. A 2-node can be any point of a line that is not a node of any other order (see Figure C). On any given line segment one can mark as many 2-nodes as one pleases.

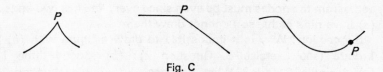

Fig. C

4.

Figure	Total number of nodes	Number of 1-nodes	Number of 3-nodes	Number of 4-nodes	Number of 5-nodes
(a)	5	0	4	1	0
(b)	2	2	0	0	0
(c)	6	5	0	0	1
(d)	5	0	0	1	4
(e)	20	7	13	0	0

✳ Note that 2-nodes have not been included in the table, since every point of an arc is a 2-node with the exception of its ends.

5. *l, s, u, v, w* and *z* (as printed in the text) are topologically equivalent to the letter *c*. Each has two 1-nodes. This question draws attention to the fact that equivalent figures must have the same number and order of nodes. See also Exercise E, Question 4.

6. *h, n, r* and *t* (as printed in the text) are topologically equivalent to the letter *y*. Each has three 1-nodes and one 3-node.

Topology

Investigation 1

Some possible solutions are shown in Figure D.

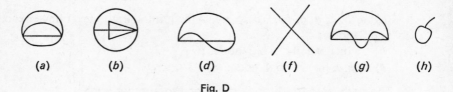

(a)　　　(b)　　　(d)　　　(f)　　　(g)　　　(h)

Fig. D

It is impossible to draw a figure which has an odd number of odd nodes ((c) and (e)).

Pupils may state this rule in other forms, such as: 'The total number of paths from the nodes must be even since every line has two ends. (c) and (e) have nine paths, so I cannot draw them.'

Suggestion: Why is it impossible to draw a figure with (a) a single 1-node (see Exercise B, Question 2); (b) exactly three 1-nodes; (c) a single 3-node? What do these impossible situations have in common?

3. ARCS AND REGIONS

Note that an arc can join a node to itself. The pupils' text states that a line joining two nodes is an arc; the two nodes need not be distinct.

It is a common mistake to forget that the whole of the outside of a figure constitutes a single region. It might help to regard this as a region bounded by the outside of the figure and the edges of the page.

Investigation 2

(a)

Figure	N	A	R
(a)	3	6	5
(b)	5	10	7
(c)	5	8	5
(d)	7	7	2
(e)	4	6	4
(f)	2	3	3
(g)	2	4	4
(h)	12	19	9
(i)	7	13	8
(j)	2	2	2
(k)	8	12	6

Nodes

Investigation 1

Draw, if possible, figures which have:

	1-nodes	3-nodes	4-nodes	5-nodes
(a)	—	—	2	—
(b)	—	1	1	1
(c)	—	3	—	—
(d)	—	2	1	—
(e)	1	1	—	1
(f)	4	—	1	—
(g)	—	2	2	—
(h)	1	1	—	—

Make up some more examples of your own. When is it impossible to draw a figure? Try to find a rule for deciding whether or not a figure can be drawn.

3. ARCS AND REGIONS

A line joining two nodes is an *arc*. An area bounded by arcs is a *region*. The area outside a figure is also a region. Each of the networks in Figure 15 has 4 nodes, 6 arcs and 4 regions. (2-nodes are not counted. Why?)

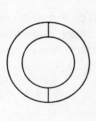 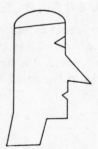

Fig. 15

Investigation 2

(a) Using some of the networks in Figure 16 and others of your own, make a table showing the number of nodes (N), arcs (A) and regions (R) in each of the networks. (Do not count 2-nodes.)

(b) Ask your neighbour to check your results.

(c) Look for patterns in your table. Comment on these patterns.

Topology

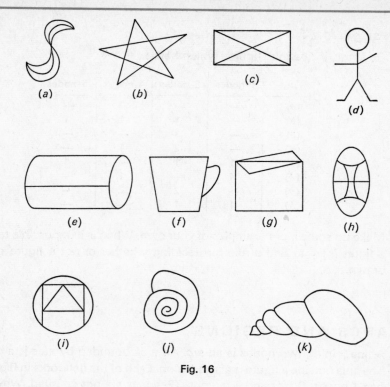

Fig. 16

4. TRAVERSABLE NETWORKS

A figure is said to be *traversable* if it can be drawn with one sweep of the pencil, without lifting the pencil from the paper, and without tracing the same arc twice. It is permitted to pass through nodes several times. The networks in Figure 17 are traversable.

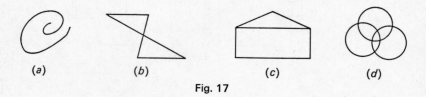

Fig. 17

Exercise C

1   Make rough sketches of the drawings in Figure 17 to show that they are traversable. Mark the point where you start (*S*) and the point where you finish (*F*).

Arcs and regions

(*b*) Some pupils draw complicated networks of their own and errors in counting can easily arise. It is, therefore, important to check entries in the table as even one error can impede the search for patterns.

(*c*) Probably the most important pattern which can be seen in all the cases tested is $N+R = A+2$. This should remind pupils of $F+V = E+2$ (*F, V, E* being the number of faces, vertices and edges respectively), Euler's formula for a polyhedron. The reason for this is made clear when the network is thought of as being formed from a polyhedron by removing a face, pulling the edges of this face outwards (remembering that they will stretch!) and pinning the whole figure flat onto a board.

The authors have deliberately refrained from giving a lead for the finding of Euler's network formula, such as suggesting that an additional column for $N+R$ be added, since this directs pupils' attention towards one particular form of one particular formula. It is possible that the result will be found in many forms, such as:

$N-A+R = 2$, $N-2 = A-R$, $R-2 = A-N$, $A+2-N = R$, etc.

This can be followed by a useful discussion as to whether or not these formulae are equivalent. With less able pupils, different hints could be given to different groups thus contriving a varied set of results, but perhaps it should be remembered that no work on equations has been attempted in this course so far.

Topology

Other results which have been noted by pupils include:
 (i) $N+A+R$ is always even;
 (ii) $A+R-N$ is always even;
 (iii) $A \times R \times N$ is always even;
 (iv) at least one of N, A, R must be even;
 (v) either (a) N, A, R, are all even, or (b) two are odd and one is even.

It should be noted that the formula $N+R = A+2$ remains true when a finite number of 2-nodes are counted. In general, 2-nodes may be ignored; for example, unless further arcs join onto the vertices, a square may be regarded as a single arc with one 2-node on it (see Figure E). Pupils are told not to count 2-nodes, but they may produce figures of their own, such as a square or a circle, where they will need to do so.

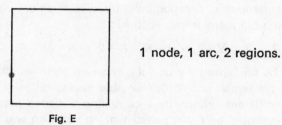

1 node, 1 arc, 2 regions.

Fig. E

Suggestion: Make an additional column for, for example, $N+R$.

4. TRAVERSABLE NETWORKS
Exercise C

With more able pupils, this exercise could be omitted; the ideas involved will arise in investigation 3.

1 Some possible solutions are shown in Figure F.

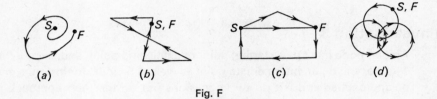

Fig. F

(*a*) and (*c*) S and F can be interchanged.
(*b*) and (*d*) S can be at any point on the network; S and F must be the same point.

Topology

3 All networks with two 1-nodes (and no other nodes) are traversable. The start must be at one of the nodes and the finish at the other.

4 All networks with two 4-nodes (and no other nodes) are traversable. The start can be at any point on the network; the start and finish must be the same point.

5 Traversable (a) (b), (f), (g), (i), (j).
   Not traversable (c), (d), (e), (h), (k).

Investigation 3

An odd node must be a starting point or a finishing point, and, except for a 1-node, will be an intermediate point as well. A 5-node, for instance, may be approached and left on two occasions but on the third approach no exits remain unused and the node must be a finishing point. Since any traverse has a start and a finish just two odd nodes are possible. Alternatively there may be no odd nodes, in which case the traverse starts and finishes at the same point; this point can be anywhere on the network.
Suggestion: Look carefully at your answers to Exercise C.

Investigation 4

(a) 3 strokes;   (b) 2 strokes;   (c) 10 strokes;
(d) 2 strokes;   (e) 4 strokes.

The number of strokes required when another line is added to Figure 18(d) depends on which extra line is drawn (see Figure G).

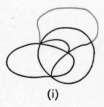
(i)
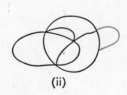
(ii)
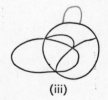
(iii)

Fig. G

Figure (i) is traversable since the number of odd nodes is reduced from four to two; Figure (ii) needs two strokes since the number of odd nodes is unchanged; Figure (iii) needs three strokes since the number of odd nodes is increased from four to six.

Traversable networks

2 Draw three traversable networks of your own, showing the start and finish.

3 Draw a traversable network with two 1-nodes and no other nodes. Where can you start and finish? Can you start from more than one point? Can you draw a network with just two 1-nodes which is not traversable?

4 Draw a traversable network with two 4-nodes and no other nodes. Where can you start and finish?

5 Which of the networks in Figure 16 are traversable? You may have to try several starting points before you either succeed or decide that the network is not traversable.

Investigation 3

Investigate possible starting and finishing points in traversable networks. When is a network traversable?

Investigation 4

The drawings in Figure 18 are not traversable. What is the least number of strokes in which each can be drawn? Compare your answers with those of your neighbour.

Add another line to Figure 18 (*d*). Do you now need fewer strokes, more strokes, or the same number of strokes? Does your answer depend on which extra line you draw?

How many extra lines must you add to Figure 18 (*a*) in order to make it traversable?

Draw other figures of your own. What is the least number of strokes in which each can be drawn? Can you find a rule for deciding how many strokes you need by just looking at a figure?

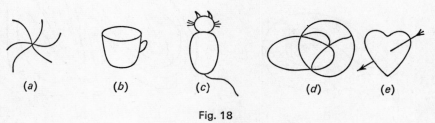

(*a*)   (*b*)   (*c*)   (*d*)   (*e*)

Fig. 18

Topology

5. INSIDE OR OUTSIDE

Fig. 19

Investigation 5

(a) Figure 20 shows a simple closed curve and two points A and B. Which point is outside the curve? Can you reach B from A without crossing the curve?

(b) Figure 21 also shows a simple closed curve and four points W, X, Y, Z. Which of the four points are inside the curve? How can you tell?

(c) Draw other simple closed curves. There is one in Figure 19. Try to find a rule for deciding which points are inside your curves.

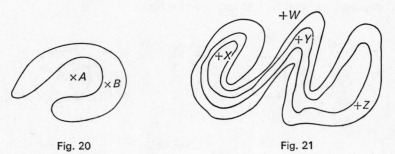

Fig. 20   Fig. 21

Traversable networks

Two extra lines which reduce the number of odd nodes from six to two will make Figure 18(a) traversable. See Figure H.

Fig. H

If a figure has n odd nodes, then the least number of strokes in which it can be drawn is $\frac{1}{2}n$. Remember that it is impossible to draw a figure with an odd number of odd nodes, so n will always be even. A figure which has no odd nodes is traversable and can therefore be drawn with one stroke.

Thus a figure with n odd nodes ($n \neq 0$) requires $\frac{1}{2}n$ strokes.

Suggestion: Make a table for the figures you have drawn showing the number of odd nodes and the least number of strokes required.

Topology

5. INSIDE OR OUTSIDE?

This section refers to simple closed curves; the results do not apply to curves in general.

Investigation 5

(a) *A*. Any line in the plane of the paper that starts at *B* and finishes at *A* must cut the curve.

(b) *X* and *Z*. Two points are on the same side if they can be joined by an unbroken curved line; a point on the outside can be joined to a point at infinity by an unbroken curved line, whereas a point on the inside cannot (we assume the curve is finite).

Some pupils may solve the inside or outside problem—even in the more difficult cases—by colouring. They start with a part of the figure which they are convinced is outside (or inside) and continue from there. Eventually the 'landmass' is separated from the 'sea' and the problem is solved (see Figure I). Lengthy, but effective!

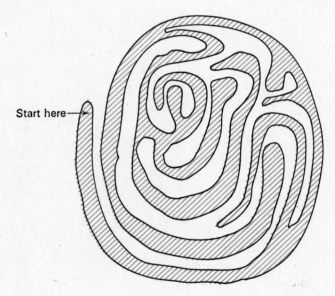

Fig. I

Inside or outside?

(c) A test which is simpler to use than that described above is hinted at in part (a). A straight line segment that starts outside and finishes inside must cut the curve in an odd number of points (see Figure J).

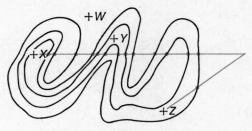

Fig. J

Suggestion: Draw several different simple closed curves. Colour the outside. Choose a point inside each curve and draw a straight line from your chosen point to the outside of the curve. Count how many times the line cuts the curve. Can you now discover a rule?

Topology

6. COLOURING REGIONS

(*a*) 3 colours; 4 colours; 2 colours; 3 colours; 4 colours. See Figure K where *A, B, C, D* stand for different colours.

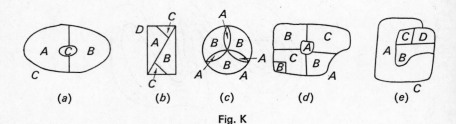

Fig. K

(*b*) Pupils enjoy being challenged to colour a given map in fewer colours.

(*c*) A large number of maps will probably be suggested and it will be necessary to demonstrate that each of these requires no more than four colours. This is sometimes hard to do!

(*d*) One possible map is shown in Figure L.

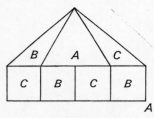

Fig. L

Exercise D

1 See Figure M. Lettering with the same letter areas which are to be the same colour is as efficient as using coloured pencils, though most pupils find this method less enjoyable.

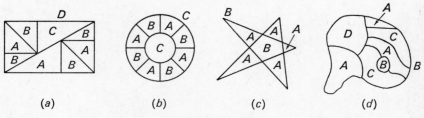

Fig. M

6. COLOURING REGIONS

(a) Copy the networks in Figure 22. Colour them so that regions with a common arc have different colours. Do not forget the outside region! (Regions with a common node may have the same colour, so long as they do not also have an arc in common.) Try to use as few colours as possible. State the number of colours you need in each case.

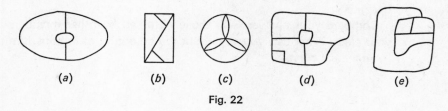

(a)   (b)   (c)   (d)   (e)

Fig. 22

(b) Draw networks of your own which need:
   (i) only two colours;   (iii) at least four colours.
   (ii) at least three colours;

Can your neighbour colour any of your networks with fewer colours than you have used?

(c) Can you find a map that needs at least five colours?

(d) Draw a map with eight regions which needs only three colours.

No design has yet been found which needs more than four colours. It has been proved that no more than five colours are needed; it seems that four will do, but this is a result which has yet to be proved—or disproved!

If you think you have found a map which needs five colours, see if some-one else can colour it in four. If not, you will have made mathematical history.

Exercise D

1 Copy the patterns in Figure 23 and colour them using as few colours as possible.

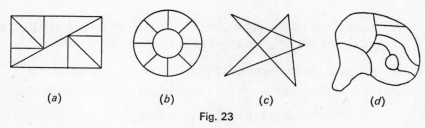

(a)   (b)   (c)   (d)

Fig. 23

Topology

2 Trace a map of northern or southern England showing the counties, and colour it with four colours. How many are usually used in an atlas?

3 Draw a unicursal curve, that is, a continuous line which comes back to its starting point. It may cross itself as often as you like, but you must not re-trace an arc already drawn (see Figure 24).

How many colours do you need to colour the regions of your design?

Compare your answer with those obtained by other members of your class. What can you say about the colouring of regions formed by unicursal curves?

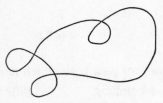

Fig. 24

4 Draw the tessellations shown in Figure 25 and colour them using as few colours as possible.

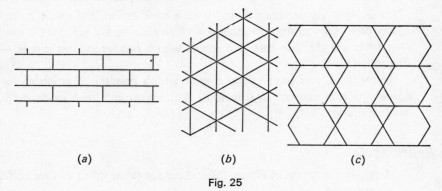

(a)   (b)   (c)

Fig. 25

5 Draw some of the eight semi-regular tessellations (see Chapter 2, p. 21) and colour them using as few colours as possible.

6 Design repeating patterns of your own which are suitable for:
(a) a kitchen floor;   (b) a book jacket;   (c) a wallpaper.

Colour them using as few colours as possible.

Colouring regions

2 Five with blue for the sea is common; they do not need so many.

3 Any design formed by a unicursal curve can be coloured in two colours. See, for example, Figure N.

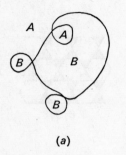
(a)

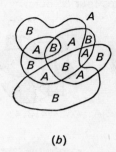
(b)

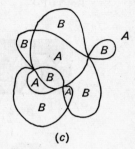
(c)

Fig. N

4 See Figure O.
  (a) Three colours are needed;   (b) two colours are needed;
  (c) two colours are needed.

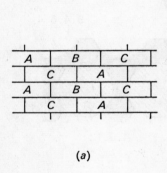
(a)

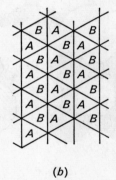
(b)

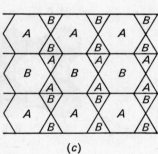
(c)

Fig. O

Topology

5 See Figure P. There are other solutions for the patterns $3^3.4^2$, $3^2.4.3.4$, and $3^4.6$.

The patterns $3.6.3.6$ and $3.4.6.4$ require two colours; the patterns $3^3.4^2$, $3^2.4.3.4$, $3^4.6$, 4.8^2 and $4.6.12$ require three colours; the pattern 3.12^2 requires four colours.

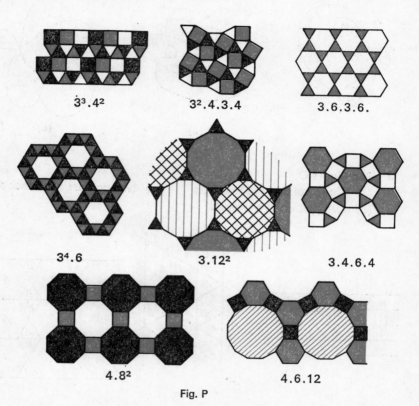

Fig. P

Colouring regions

Summary

A *topological transformation* allows twisting and stretching but not tearing or joining.

An *invariant* under a topological transformation is a fact about the original figure which is still true about the new one.

Two curves are topologically *equivalent* if one can be transformed into the other by a topological transformation.

A *simple closed curve* is equivalent to a circle.

A *node* is a point with at least one path leading from it. The order of a node is the number of paths.

A line joining two nodes is an *arc*. Every point of an arc is a 2-node with the exception of its ends.

An area bounded by arcs is a *region*. The area outside a figure is also a region.

A figure is *traversable* if it can be drawn with one sweep of the pencil, without lifting it from the paper and without tracing the same arc twice.

Here is a list of the invariant facts we have already met. (Dashed letters are used for the points and lines into which the original ones are twisted and stretched.)

(a) If a point P lies on a line l, then P' lies on l'.

(b) If P_1, P_2, P_3 lie in that order on l, then P'_1, P'_2, P'_3 lie in that order on l'.

(c) If there are n paths from P, then there are n paths from P'. In Figure 26 (c), there are eight paths from P.

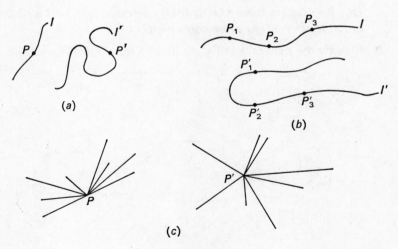

Fig. 26

Topology

It is not possible to draw a figure which has an odd number of odd nodes.

For any figure, $N+R = A+2$, where N = number of nodes, R = number of regions and A = number of arcs.

A figure is traversable if it has: (*a*) two odd nodes, (*b*) no odd nodes. It may have any number of even nodes.

A line segment that starts outside a simple closed curve and finishes inside cuts the curve in an odd number of points.

Networks are coloured so that regions with a common arc have different colours. No one has yet found a plane design which needs more than four colours.

Exercise E (Miscellaneous)

1   Which of the following are simple closed curves:
   - (*a*) an equilateral triangle;
   - (*b*) an octagon;
   - (*c*) a letter *N*;
   - (*d*) a figure 8;
   - (*e*) a letter *R*?

2   Would you consider 'inside' and 'outside' to be topological invariants?

3   The parallelogram *ABCD* is transformed into the square *EFGH*. Is it always, sometimes or never true that:
   - (*a*) $A \to E$, $B \to F$, $C \to G$, $D \to H$;
   - (*b*) $A \to E$, $B \to G$, $C \to F$, $D \to H$;
   - (*c*) $A \to G$, $B \to F$, $C \to E$, $D \to H$?

4   (*a*) Two figures are equivalent. Must they have the same number and order of nodes?

   (*b*) Two figures have exactly the same number of nodes of the same order. Must the figures be equivalent?

5   Classify the networks in Figure 27 into topologically equivalent sets.

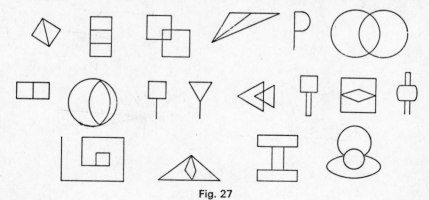

Fig. 27

Colouring regions

Exercise E

In general, the questions in this exercise are harder than those in exercises A, B, C, D.

1 (a) Yes;   (b) yes;   (c) no;   (d) no;   (e) no.

2 In two dimensions, insides and outsides remain insides and outsides.

3 This question stresses that, although order is preserved, it is impossible to know the position of a 2-node after the transformation.
   (a) sometimes true;   (b) never true;
   (c) sometimes true if the figure may be turned over.

4 (a) If two figures are equivalent then they have the same number of nodes of the same order. This is a necessary, but not a sufficient, condition for equivalence. Figure Q shows two figures which are not equivalent although each has just two 4-nodes. In Book C we shall see that the matrices associated with these networks are different.

Fig. Q

Topology

5 The networks can be divided into five sets as follows:

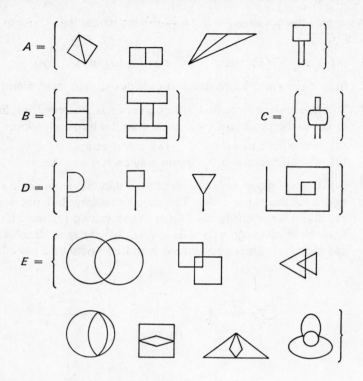

6 Figure R shows a possible sketch.

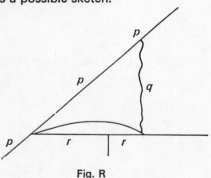

Fig. R

There are two cross-roads, provided a cross-road is defined as any junction of four roads. The figure shows that the roads are not opposite each other. There are two or four junctions according to your decision about the nature of a cross-road. The questions are intended to provoke discussion, there are no required answers.

Colouring regions

6   Figure 28 shows a topological map of some roads. How many cross-roads are there? How many junctions are there?

The three sections marked *p* actually form a straight road running SW–NE. The section *q* is winding but runs roughly N–S. The sections *r* form a straight road running E–W. Sketch your idea of what the map really looks like. Mark the *p*'s, *q*'s and *r*'s on your sketch.

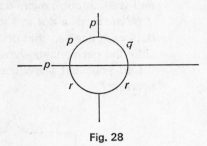

Fig. 28

7

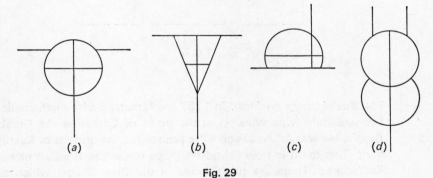

Fig. 29

Look carefully at the drawings in Figure 29. Which of them are possible topological maps of the roads shown in Figure 30?

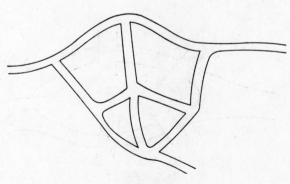

Fig. 30

Topology

8   Figure 31 shows a plan view of a small bungalow. Can you start at *A* and walk through every door of the bungalow exactly once?

(*Hint*: mark a dot in each room and a single dot outside, join the dots with lines so that one line passes through each door, and consider the network you have now drawn.)

Can you start anywhere except *A* and succeed in traversing the network?

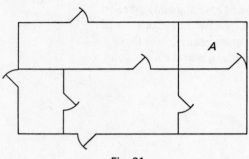

Fig. 31

9   *The Koenigsberg problem.* In 1737 the famous Swiss mathematician Leonard Euler was working at the court of Catherine the Great of Russia. He was asked to solve the problem of the bridges of Koenigsberg. This town is now found on maps under the Russian name of Kaliningrad. There are two islands in the River Pregel which runs through the town. Seven bridges cross the river as shown in Figure 32.

The problem was this: is it possible to take a walk which crosses each of the bridges once and once only? Euler solved the problem by changing it into a problem about the nodes of a network. See if you can solve the problem.

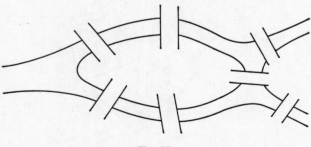

Fig. 32

Colouring regions

7 All of them.

8 Figure S shows the house reduced to a network which has one 4-node, one 3-node and one 1-node. Since the network has two odd nodes, the route is traversable. It is also possible to start outside the house, at the 3-node, but nowhere else.

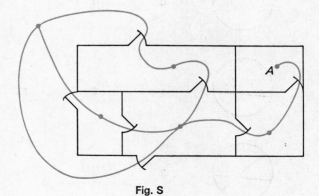

Fig. S

9 The network for the Koenigsberg Bridge problem is shown in Figure T. It has three 3-nodes and one 5-node and is therefore not traversable.

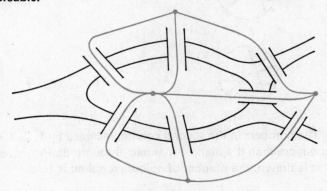

Fig. T

Topology

10 The table is given below with a sketch of the circle at each stage.

Figure	Number of lines	Number of regions	Number of colours
	0	1	1
	1	2	2
	2	4	2
	3	7	2
	4	11	2

The numbers in the second column increase by 1, 2, 3, 4, ... and are one more than the triangle numbers throughout. Provided at least one line is drawn, the number of colours required is two.

Colouring regions

Investigation 6

Figure U shows pentagons with 0, 1, 2, 3, 5 crossings; it is impossible to draw a pentagon with 4 crossings.

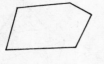

Fig. U

Topology

Figure V shows hexagons with 0, 1, 2, 3, 4, 5, 6, 7 crossings.

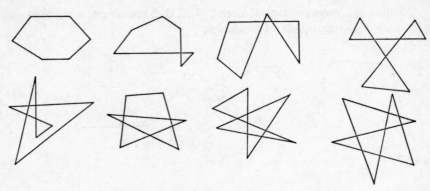

Fig. V

A pair of adjacent sides of a polygon meet at a vertex and therefore cannot provide a crossing; a pair of non-adjacent sides cannot provide more than one crossing. An n-sided polygon has $\frac{1}{2}\{n(n-3)\}$ pairs of non-adjacent sides. It is therefore impossible for the polygon to have more than $\frac{1}{2}\{n(n-3)\}$ crossings. If n is odd, polygons can be drawn with any number of crossings from 0 to $\frac{1}{2}\{n(n-3)\}$ inclusive with the exception of $\frac{1}{2}\{n(n-3)\}-1$. If n is even, then it is possible to draw polygons with any number of crossings from 0 to $\frac{1}{2}\{n(n-4)\}+1$ inclusive. (The proof of these facts would occupy more space than is justified in this book.) The results are stated for the benefit of the teacher; it is not expected that they will be produced by pupils.

The investigation has been included because the problem posed is easy to understand and it is hoped that pupils will be stimulated to ask other questions, such as: 'Are polygons unicursal doodles?', 'What is the inside of a polygon?', 'Is it possible to have triple intersection?'

Colouring regions

10 Draw a circle of radius 5 cm. Draw a straight line from one point on the circle to another (see Figure 33 (a)). There are now two inside regions and the inside can be coloured with two colours.

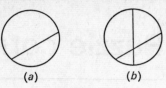

(a)   (b)

Fig. 33

Draw a second straight line so as to make as many new regions as possible (see Figure 33 (b)). How many regions are there? How many colours are now needed to colour the interior? Draw a third straight line to make as many new regions as you can. Continue the process and enter your results in a copy of the following table:

Number of lines	Number of regions	Number of colours
0	1	1
1	2	2
2		
3		
.		
.		
.		

Look for patterns in your table and comment on those that you find.

11 *The colour game.* This is a game for two players. The first player draws a region. The second player colours it and draws a new region. The first player colours this region and adds a third. The game continues until one of the players is forced to use a fifth colour; this player is the loser. Play this game with a friend.

Investigation 6

Figure 34 shows a 'crossed-over' pentagon with one one crossing. Can you draw crossed-over pentagons with 2, 3, 4, 5 crossings? What can you say about crossed-over hexagons? Investigate the numbers of crossings which are possible in crossed-over polygons.

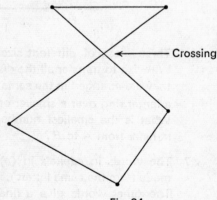

Crossing

Fig. 34

Puzzle corner

1.
```
  CROSS
+ ROADS
-------
 DANGER
```

Each letter stands for a number. If O represents 2 and S represents 3, what do the other letters represent?

2. In how many different ways can a pair of dice land?

3.
```
    BC
  × BC
  ----
   ABC
```

What numbers do A, B and C represent?

4. How can 6 identical match sticks be arranged to form 4 equilateral triangles?

5. Among twelve similar coins there is one 'fake' which weighs less than the others. Explain how, using only three weighings of a balance, you can find the fake coin.

6.
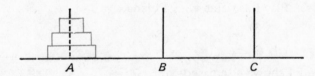

Three discs of different sizes are arranged as shown in the figure. We wish to transfer all the discs, one at a time, on to spike B, so that they are arranged in the same way. (At no time is it permitted to place a larger ring over a smaller one, but you can of course use spike C.) What is the smallest number of moves necessary to complete the transfer from A to B?

7. The words in capitals in (a)–(d) form an anagram (another word made from the same letters), in each case a single mathematical word. The other words give a final clue. For example, 'LARGE TIN with three sides' would be TRIANGLE.

Puzzle corner

1. Solution by trial and error.

   96233 $R = 6, D = 1, E = 4, A = 5, G = 7, N = 8, C = 9.$
   +62513
   ──────
   158746

2. 36 ways.
   Imagine a red and a blue die. Then any one of the 6 numbers on the red die can be combined with any one of the 6 numbers on the blue die.

3. Solution by trial and error.

   25       $A = 6, B = 2, C = 5.$
   ×25
   ───
   625

4. The match sticks must be arranged to form a regular tetrahedron (see Figure A).

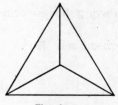

Fig. A

5. Step 1. Place 6 coins on each pan of the balance. The 'fake' must be among the 6 coins that weighed the least.
   Step 2. Of these 6 coins, place three on each pan. Choose the 3 coins weighing the least.
   Step 3. Place one of these on each side of the balance. To find the fake coin is now straightforward.

Puzzle corner

6   This is a very simple version of the 'Tower of Hanoi' problem. (See *Mathematical Recreations and Essays,* by Rouse-Ball.)
    The transfer of the discs to spike *B* can be completed with a minimum of 7 moves. If we label the discs *x, y, z* respectively, *x* being the smallest one, then the 7 moves are as follows:

$$x \to B, \quad y \to C, \quad x \to C, \quad z \to B, \quad x \to A, \quad y \to B, \quad x \to B.$$

Pupils might find it easier to consider discs of different colours.

7   (*a*) Binary;   (*b*) Subtraction;
    (*c*) Coordinate;   (*d*) Reflection.

8   See Figure B.

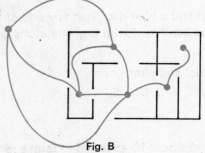

Fig. B

The topological drawing has 4 odd nodes and is therefore not traversable.
The answer to the question is 'No'.

T316

Puzzle corner

9 See Figure C.
   (Reference: Rouse-Ball, *Mathematical Recreations and Essays*.)

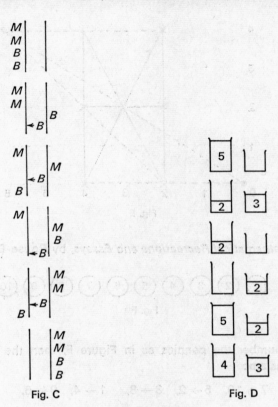

Fig. C

Fig. D

10 See Figure D.

Puzzle corner

11 The extra points have coordinates (0, 0) and (6, 4) (see Figure E).

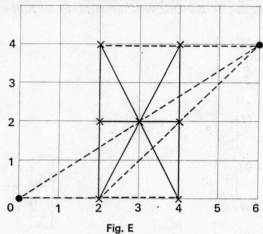

Fig. E

12 See *Mathematical Recreations and Essays,* by Rouse-Ball.

① ② ③ ④ ⑤ ⑥ ⑦ ⑧ ⑨ ⑩

Fig. F

If we number the pennies as in Figure F, then the two possible solutions are:

$\quad\quad\quad\quad 7 \to 10, \quad 5 \to 2, \quad 3 \to 8, \quad 1 \to 4, \quad 9 \to 6,$

or $\quad\quad 4 \to 1, \quad 6 \to 9, \quad 8 \to 3, \quad 10 \to 7, \quad 2 \to 5.$

13 Solution by trial and error.

$$\tfrac{3}{6} = \tfrac{7}{14} = \tfrac{29}{58} \quad \text{or} \quad \tfrac{3}{6} = \tfrac{9}{18} = \tfrac{27}{64} \quad \text{or} \quad \tfrac{2}{6} = \tfrac{3}{9} = \tfrac{58}{174}.$$

Puzzle corner

(a) BRAINY; a system of counting numbers.
(b) NO CAR—IT BUST; we certainly find a difference.
(c) I DO NOT CARE, and this helps to make the position clear.
(d) NO RICE LEFT: that's the effect of a mirror.

8 The diagram shows a simple ground floor plan of a house. Could you walk through this house so that you walked through each door once and once only?
   (*Hint*: make a topological drawing by making each room into a node and each door into an arc.)

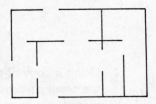

9 Two men and 2 boys want to cross a river but they only have a canoe which will carry 1 man or 2 boys. How do they get across?
   (*Hint*: the figure shows the first stage.)

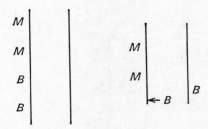

10 A man goes to a barrel with 2 jars; one holds 3 litres and the other 5 litres. Explain how he can measure 4 litres.

11 Plot the points (2, 0), (2, 2), (2, 4), (3, 2), (4, 0), (4, 2), and (4, 4). Draw five line segments each containing three of these seven points.
   Now find two more points which will enable you to draw four more line segments each containing three of the new total of nine points.

12 Ten pennies are placed in a row touching one another. Any penny may be moved over *two* of those next to it on to the coin beyond. How can you move the coins in this way so that they will be arranged in equally spaced pairs?
   (*Hint*: move only those coins in odd positions *or* only those in even positions.)

13 Use all the numbers from 1 to 9 once each to form three equivalent fractions as in the following example:

$$\tfrac{2}{4} = \tfrac{3}{6} = \tfrac{79}{158}.$$

Puzzle corner

14

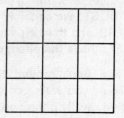

Using a copy of the figure, arrange the set of numbers from 1 to 9 in the small squares, called cells, so that if you add the numbers written in any row or column, or diagonal, you will always obtain the same answer. (Square arrays of numbers with these properties are known as 'Magic Squares' and were known to the Chinese over 2000 years ago.)

Puzzle corner

14 Basically this is the only pattern for a 3 × 3 square. All others are reflections and/or rotations of this one, making a total of 8.

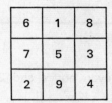

Fig. G

Some pupils might like the hint that the sum of numbers in any row, column or diagonal is 15.

Revision exercises

Quick quiz, no. 3

1. The figure has 4 even nodes and 2 odd nodes and is therefore traversable.
2. $x = 3$.
3. (a) 11_{two}; (b) 345_{six}; (c) 777_{eight}.
4. (a) £9·15; (b) $2·83.
5. (a) 240°; (b) 320°; (c) 290°; (d) 110°.

Revision exercises

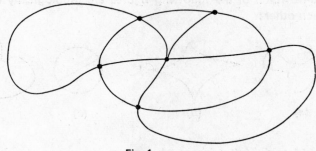

Fig. 1

Quick quiz, no. 3

1. Is Figure 1 traversable?
2. A line joins (3, 0) to (3, 6). What is its equation?
3. Subtract 1 from the following numbers:
   (a) 100_{two}; (b) 350_{six}; (c) 1000_{eight}.
4. Work out the following:

   (a) £
       2·37
       5·88
   + 0·90
   ────

   (b) $
       4·82
   − 1·99
   ────

5. What are the bearings of
   (a) A from B;
   (b) B from C;
   (c) A from C;
   (d) C from A?

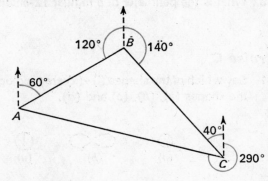

T323 147

Revision exercises

6 On to what set is {2, 4, 6, 8} mapped by the following:
   (a) $x \to 3x$;   (b) $x \to \frac{1}{2}x$?

Quick quiz, no. 4

1 State which of the following figures are topologically equivalent to each other:

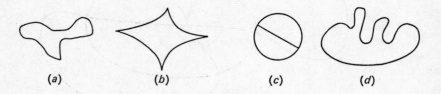

2 Add 1 to the following numbers:
   (a) 1011_{two};   (b) 102_{three};   (c) 144_{five}.

3 Give a possible value for x if
   (a) $x > {}^-2$;   (b) $x < {}^+1$;
   (c) $^-4 < x < 0$;   (d) $^-7 < x < {}^-2$.

4 What is the final bearing after
   (a) start facing SE and do an anticlockwise turn of 110°;
   (b) start facing NW and do an anticlockwise turn of 280°;
   (c) turning 170° clockwise from east?

5 What mapping in the form $x \to$? when applied to {2, 4, 6, 8} maps it on to
   (a) {5, 7, 9, 11};   (b) {1, 3, 5, 7}?

6 What is the perimeter of a regular 12-sided polygon of side 2·5 cm?

Exercise C

1 Say which of the shapes (i)–(x) are topologically equivalent to each of the shapes (a), (b), (c) and (d).

148   T324

Revision exercises

6 (a) $\{6, 12, 18, 24\}$;   (b) $\{1, 2, 3, 4\}$.

Quick quiz, no. 4

1 (a), (b) and (d) are topologically equivalent to one another.
2 (a) 1100_{two};   (b) 110_{three};   (c) 200_{five}.
4 (a) $025°$;   (b) $035°$;   (c) $260°$.
5 (a) $x \to x+3$;   (b) $x \to x-1$.
6 30 cm.

Exercise C

1 (i), (vii), (ix) are topologically equivalent to (a);
(iii), (x) are topologically equivalent to (b);
(ii), (iv), (vi), (viii) are topologically equivalent to (c);
(v) is topologically equivalent to (d).

Revision exercises

2   (a) 10001001;   (b) 101111;   (c) 11101010;
    (d) 1101111;   (e) 1010110.

3   (a) (i) ⁺4;   (ii) ⁺7.40;   (iii) ⁻3.48;   (iv) ⁻6.13.
    (b) (i) 11.47 h;   (ii) 21.08 h;   (iii) 19.39 h;   (iv) 0.42 h.

4   090°, 150°, 210°, 270°, 330°, 030° (assuming a *regular* hexagonal course).

5

	No. of degrees
Hotel, boarding house, etc.	75
Holiday Camp	$43\frac{1}{2}$
At home or staying with friends	$115\frac{1}{2}$
Caravan	27
Tent	45
Boat	$37\frac{1}{2}$
Others	$16\frac{1}{2}$
	360

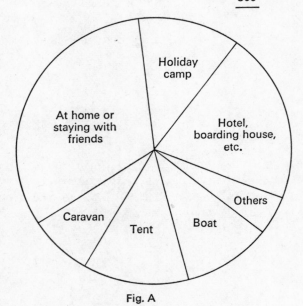

Fig. A

Revision exercises

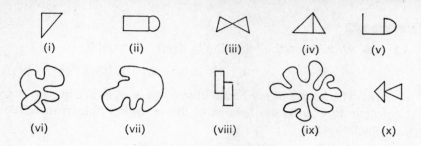

2 Combine the following binary numbers:
   (a) 101101 + 1011100;   (b) 1011100 − 101101;
   (c) 10010 × 1101;   (d) 110110 + 101010 + 1111;
   (e) 111001 + 100111 − 1010.

3 Suppose the British and continental (24-hour) systems for recording the time are to be replaced by a new system in which 12 noon becomes 'zero-hour' from which all times are measured, e.g. 2.30 p.m. (or 14.30 h) would be written as ⁺2.30 and 7.15 a.m. would be ⁻4.45.
   (a) Convert the following to the new system:
   (i) 16.00 h;   (ii) 19.40 h;   (iii) 8.12 a.m.;   (iv) 5.47 a.m.
   (b) Convert the following 'new' times to the continental system.
   (i) ⁻0.13;   (ii) ⁺9.08;   (iii) ⁺7.39;   (iv) ⁻11.18.

4 A yacht sails round a hexagonal course in a clockwise direction ABCDEF, each 'leg' of the course AB, BC, CD, . . . being 2 kilometres long. On what bearing is it sailing on each 'leg' if the first is due east?

5 A travel agency carried out a survey on the holiday habits of students. 240 students who had not gone abroad were asked to state their main type of accommodation during their previous summer holiday. The results were as follows:

Type of accommodation	Number of students
Hotel, boarding house, etc.	50
Holiday Camp	29
At home or staying with friends	77
Caravan	18
Tent	30
Boat	25
Others	11
	240

Show this information on a pie chart.

Revision exercises

Exercise D

1. (a) Write down the following in descending order:

   $^-9$, $^+6$, $^-5$, $^+3$, $^-1$.

   (b) The height of my house above sea level is 320 m. Use $+$ and $-$ signs to express the heights of the following objects in relation to my house:
   - (i) a radio mast 2200 m above sea level;
   - (ii) a church 320 m above sea level;
   - (iii) a bridge 210 m above sea level;
   - (iv) a submarine 100 m below sea level.

2. Draw an arrow diagram to show the relation 'is a factor of' between members of {2, 3, 4, 5} and members of {4, 6, 8, 10}.

   If the direction of the arrows in your diagram was reversed, what relation would your diagram show?

   Find a different relation between these two sets. (Write the answer in the form $x \rightarrow$?) Is this relation a mapping?

3. Sketch, if possible, figures with:
   - (a) One 4-node, one 3-node and one 1-node;
   - (b) one 5-node, two 3-nodes and one 1-node;
   - (c) one 1-node, one 5-node and one 4-node.

4. A girl is sitting on a beach. The entrance to the pier is 200 m away on a bearing of 310° and the end of the pier is 220 m away on a bearing of 337°. Find by accurate drawing the length of the pier.

   A boat is moored 300 m away from the girl on a bearing of 040°. How far would someone have to swim from the end of the pier to reach the boat?

5. In this question,

   $\square = 10_{ten}$ $\boxed{1} = 11_{ten}$ $\boxed{2} = 12_{ten}$

   $\boxed{3} = 13_{ten}$ etc.

   Find what number base is being used in each of the following calculations:

   (a)   1 6 $\square$ (b)   338   (c) $\square$ $\boxed{3}$ 3   (d)   4 0 2
       +3 3 3   −287   +9 1 6   −1 6 3
       4 $\square$ 0   41   1 3 $\boxed{4}$ 9   2 4 $\square$

Revision exercises

Exercise D

1 (a) +6, +3, −1, −5, −9.
   (b) (i) 1880 m; (ii) 0 m; (iii) −110 m; (iv) −420 m.

2

'is a multiple of'.

$x \to 2x$, which is mapping.

3 For one example of each, see Figure B.

 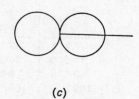

(a) (b) (c)

Fig. B

4 The length of the pier is 100 m.

The distance between the pier and the boat is 280 m.

Remember that it is important to give the scale, which on this drawing is 1 cm to 50 m.

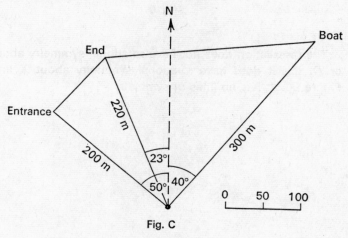

Fig. C

Revision exercises

5 (a) Thirteen;   (b) nine;   (c) sixteen;   (d) eleven.

Exercise E

2 (a)

	No. of degrees
Australia	$53\frac{1}{3}$
Canada	$56\frac{2}{3}$
France	$21\frac{2}{3}$
New Zealand	50
Italy	$61\frac{2}{3}$
U.K.	$43\frac{1}{3}$
U.S.A.	$73\frac{1}{3}$
	360

  (b) Discuss the advisability of 1 motif representing 2 oz of sweets.

3

	to 3 S.F.	to 2 S.F.	to 1 S.F.	to 1 dec. pl.	to 2 dec. pl.
7·191	7·19	7·2	7	7·2	7·19
10·270	10·3	10	10	10·3	10·27
0·526	0·526	0·53	0·5	0·5	0·53
109·3	109	110	100	109·3	109·30

These entries assume the original numbers were exact.

4

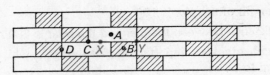

Fig. D

The tessellation does not have rotational symmetry about A, B, C or D, but it does have rotational symmetry about X and Y (see Figure D). It has no lines of symmetry.

Revision exercises

Exercise E

1. Criticize the following statements:

   (a) the distance from here to Oxford is 6·0251 km;

   (b) my car is 1·73 m wide, so if the gate posts are about 1·75 m apart I'll be able to get through;

   (c) we need 2·0371 g of butter for this cake;

   (d) you just can't trust this newspaper—there were 999 people in the audience, but they reported it as 1000.

   (e) you're supposed to dial 999 for the police, but I never do: 1000 is near enough for me!

2. The average number of ounces of sweets consumed per month by 14-year-old girls in seven different countries were once as follows: Australia 32, Canada 34, France 13, New Zealand 30, Italy 37, U.K. 26, U.S.A. 44.

   (a) Show this information on a pie chart.

   (b) Show this information by means of a pictogram.

   (c) Which method is better? Why?

   (d) Would it make any difference to your answer to (c) if all you wanted to do was to convince an unthinking English girl that she might slim more easily in France? Why?

3. Complete the following table.

	To 3 s.f.	To 2 s.f.	To 1 s.f.	To 1 dec. pl.	To 2 dec. pl.
7·191					
10·270					
0·526					
109·3					

4. Does this tessellation have rotational symmetry about any of the points A, B, C, D? Are there any (other) points about which the tessellation does have rotational symmetry? Sketch the figure and mark any such points. Mark also any lines of symmetry which the figure has.

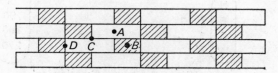

Revision exercises

5 Where necessary, complete the following arrow diagrams and state the relation that each illustrates. Each example is taken from a chapter in this book. For example:

Statistics

Method of travel   The number who use that method

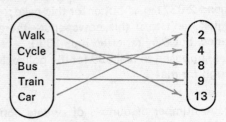

'Method of travel used by'

(a) *Relations*

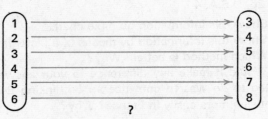

(b) *Topology*

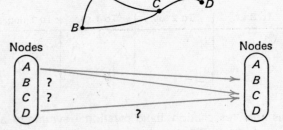

(c) *Angles*

Bearings   Clockwise rotations from north

'Gives the same direction as'.

Revision exercises

5 (a) 'Is two less than'.

(b)

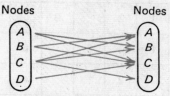

'Is on the same arc as'.

(c)

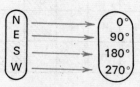

Revision exercises

Exercise F

1. 1	2. 2		3. 8	4. 9	5. 7
6. 1	0	7. 8	8. 6	9	0
	9. 2	1	10. 4	9	
11. 7					12. 4
	13. 1	14. 5	15. 1	16. 3	
17. 1	4	4	18. 2	1	19. 4
20. 2	0		21. 2	22. 5	1

Revision exercises

Exercise F

Complete this cross-number

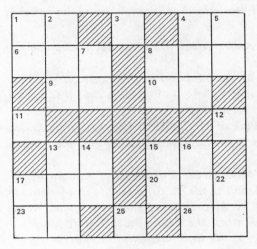

Clues across

1. The difference between the 4th and 8th prime numbers.
3. An octagon has ? lines of symmetry.
4. A 100-sided convex polygon has ? diagonals from any one vertex.
6. What is the angle (in degrees) of a regular pentagon?
8. 688 correct to 2 S.F.
9. The sixth triangle number.
10. What is the perimeter (in centimetres) of a rectangle measuring $13\frac{1}{2}$ cm by 11 cm?
11. $35 - 16 = 16$ is correct in what number base?
12. A tetrahedron has ? vertices.
13. Through how many degrees does the hour hand of a clock turn in 30 minutes?
15. If $A = \{1, 2, 3, ..., 40\}$ and $B = \{$multiples of 3 less than 60$\}$, how many members has $A \cap B$?
17. 12 dozen.
20. 214·499 correct to the nearest whole number.
23. How many fifths in four wholes?
25. A parallelogram has rotational symmetry of order ? about its centre.
26. $50\frac{6}{11}$ correct to the nearest whole number.

Clues down

1. There are ? prime numbers between 1 and 35.
2. 20_{ten} in base three.
4. The 500th positive odd number.
5. The product of the 1st, 3rd and 4th prime numbers.
7. The ninth square number.
8. 2^6.
13. 4 angles of a pentagon are each 100°. What is the fifth angle (in degrees)?
14. $64_{ten} = ?_{twelve}$.
15. A cube has ? edges.
16. Bearing of the NW direction.
17. A dodecagon has ? sides.
22. Total length (in metres) of the edges of a rectangular box measuring $2\frac{1}{4}$ m × 3 m × 5 m.

Bibliography

Budden, F. J. *An Introduction to Number Scales and Computers.* Longmans, 1965.
Coxeter, H. S. M. *Introduction to Geometry.* Wiley, 1961.
Cundy, H. M. and Rollett, A. P. *Mathematical Models* (2nd edition, 1961). Oxford University Press.
The Graphic Work of M. C. Escher (New edition). Oldbourne, 1967.
Hilbert, D. and Cohn-Vossen. *Geometry and the Imagination.* Chelsea, N.Y., 1952.
Lewis, K. and Ward, H. *Starting Statistics.* Longmans, 1969.
Pearcy, J. F. F. and Lewis, K. *Experiments in Mathematics.* Longmans, 1966.
Rouse-Ball, W. W. *Mathematical Recreations and Essays.* MacMillan, 1959.
Steinhaus, H. *Mathematical Snapshots.* Oxford University Press, 1960.
S.M.P. Handbook, *We Built our own Computers.*
Wells, A. F. *The Third Dimension in Chemistry.* Oxford University Press, 1956.

Index

(Note: references are to the red T page numbers)

abacus, addition and subtraction of decimals on, 64–9; base twelve on, 205–6; base two on, 213, 226–7; decimals on, 56, 60–1
accuracy, 54
addition, in base twelve, 208–11; of decimals on abacus, 64–9; of fractions, 133
angle, 152–79; in tessellations, 28, 30, 40, 41
arcs, 286–90, 303; formula for nodes, regions, and, 286–7, 289, 304
areas, comparison of, 82–93; measurement of, 92–103; of rectangles, 102–11
arrow diagrams, 182–5
arrowhead, 5, 7

back bearings, 162
bar charts, 236–9, 241
base five, 'cross-number' in, 200, 203; multiplication in, by Napier's bones, 201–3
base twelve, 204–11
base two, 212–26; 'cross-number' in, 230–1; fractions in, 226–9
bases greater than twelve, 210–13
bearings, 153, 156–73
binary system, *see* base two
brackets, 17

centres of rotational symmetry, 43–4, 50
circles, areas of, 98, 100; for estimating areas, 82–5
class projects, on ages, 259; on heights, 263; on polyominoes and polyiamonds, 44–7; statistical, 244–9; on tiling patterns, 8
clock-ray method of fixing a position, 154–7
coinage, decimal, 68, 70–3
colouring, of regions, 298–302, 304, 313; of tessellations, 34, 36–9, 300; of tiling patterns, 7
commutative property, 16, 211
comparison, of area, 82–93; of fractions, 128–37
compass, in class-room, 157
computer tape, 225–6
computers, D.I.B.S., 222, 223; electronic, 221–5; human, 218, 220–1; P.A.C.A., 223
congruent figures, 13
coordinates, Cartesian and polar, 175
counting, distinguished from measuring, 54, 58, 59

counting numbers, letters for, 14, 17, 18, 22, 23
course, plotting a, 160–73
'cross-numbers', in base five, 200, 203; in base two, 230, 231
cubes, edges, faces, and vertices of, 24, 25; space-filling with, 50
curves, simple closed, 275, 303; topologically equivalent, 275–81, 303; unicursal, 300

data, collection and presentation of, 234–5, 237
decimal coinage, 68, 70–3
decimal point, 56
decimals, 53–77; addition and subtraction of, 64–73; in base two, 226, 229; in measurement, 53–65; rounding off, 72–7; significant figures in, 74–7
degrees, for measuring turns, 156–60
diagonals, of polygons, 26, 27
diamonds, 46, 47
directed numbers, 260–9
distributive property, 16
dodecagon, regular, 30, 33; in semi-regular tessellations, 49
dominoes, 44
duodecimal system, 204–11

edges of polyhedra, formula for faces, vertices, and, 24, 25, 289
equations, 15, 18
equilateral triangles, 4, 29; in semi-regular tessellations, 47, 49; tessellations from, 30, 31, 32; tiling patterns from, 12, 13
equivalent fractions, 120–7
Euler's formula for polyhedra, 24, 25, 289

faces of polydedra, formula for edges, vertices and, 24, 25, 289
factors, of twelve and ten, 207
family relationships, 180–7
Farey sequences, 138–41
formulae, 22–7
fraction number line, 132–7
fraction point, 56
fractions, addition of, 133; in base two, 226–9; comparison of, 128–37; equivalence of, 115; equivalent, 120–7; Farey sequences of, 138–41; graphing of,

Index

fractions *(cont.)*
  116, 118, 119; representation of, 116–21; subtraction of, 133; of units, 55–63
frequency, rectangular and triangular distributions of, 249

graphs, of fractions, 116, 118, 119; comparison of fractions by, 130, 132–3; from relations, 198–9

hexagons, crossed-over, 312, 313; regular, 5, 9, 29, 52; in semi-regular tessellations, 47, 49; tessellations from, 32, 39, 43; tiling patterns from, 8, 9, 11
hexominoes, 45–7

identity, 15
images, in mapping, 191, 192
inside or outside, 294, 296–7, 304
intersection sets, 22, 23
invariants, topological, 275, 303
irrational numbers, 135
isosceles triangle, 1, 4

kite, 1, 5, 7
Koenigsberg Bridge problem, 308, 309

length, units of, 54, 55; fractions of units of, 55–63; standard units of, 60, 62–5
letters, for numbers, 14–27; order of frequency of, 244, 247, 249
lines of symmetry, in tessellations, 44, 50

magic squares, 320, 321
mapping machine, 192, 193
mappings, 190–5
measurement, of areas, 92–103; decimals in, 53–65
metre, 54, 60
multiplication, in base twelve, 210, 211; by Napier's bones, 201–2

Napier's bones, 201–2
networks, traversable, 288, 291–3, 295, 303, 304
nodes, 282–7, 303; formula for arcs, regions, and, 286–7, 289, 304
non-numerical sets, 18–19
north line, bearings measured from, 156–9
number bases, 200–33
number line, 254, 264–9; for fractions, 132–7
number names, 79–81
number symbols, 210, 211
number systems, 132
numbers, counting, letters for, 14, 17, 18, 22, 23; directed, 260–9; irrational, 135; letters for, 14–27; patterns among, 14–19; rational, 132, 135; rectangle, 14; triangle, 25

octagon, regular, 30, 33; in semi-regular tessellations, 49
octahedra, edges, faces, and vertices of, 24, 25; regular, space-filling with, 50; truncated, 52
ordered pairs, fractions as, 116, 118, 119; representing relations as, 196–9

parallel lines, and alternate angles, for calculating bearings, 162
parallelogram, 5, 7; tiling patterns from, 11
patterns, in addition and multiplication tables, 208–11; number of ways of repeating, 28; among numbers, 14–19; in tessellations, 28–41, 43; in tiling, 1–13
pentagons, crossed-over, 311, 313; regular, 9, 29, 32, 33; tiling patterns from, 8, 9, 11
pentiamonds, 46, 47
pentominoes, 44, 45
pictograms, 242–5
pie charts, 240–2
polygons, crossed-over, 311–13; diagonals of, 26, 27; tiling patterns from, 8–12
polyhedra, edges, faces, and vertices of, 24, 25, 289; semi-regular, 52; space-filling, 50
polyiamonds, 46, 47; tessellations from, 46
polyominoes, 44, 45; tessellations from, 46, 47
position, fixing a, 152–3, 155; by bearings, 156–73; by clock-ray method, 154–7; by radar, 174–9
prisms, space-filling with, 50
punched cards, binary numbers on, 227
pyramid (square-based), edges, faces, and vertices of, 24, 25

quadrilaterals, tiling patterns from, 6–8

radar, 174–9
railway marshalling yard, as two-state system, 228
range card, 154
rational numbers, 132, 135
rectangle numbers, 14
rectangles, 1, 7; areas of, 102–11; comparison of areas by counting, 90, 91; tessellations from, 31–5
reference points, 254–60
regions, 272, 286–90, 303; colouring of, 298–302; formula for arcs, nodes, and, 286–7, 289, 304
relations, 180–7; between sets, 188–91; represented by ordered pairs, 196–9
rhombus, 1, 5; tiling patterns from, 11
right-angled triangle, 1–3
rotational symmetry, 12, 13; in tessellations, 42–4, 48, 50, 51
rounding off, 72–5
rubber-sheet geometry, 272

Index

scalene triangle, 4
semi-regular polyhedra, 52
semi-regular tessellations, 46–9
sets, 18–22; intersection of, 22–3; relations between, 188–91
shifts, game of, 252–3
significant figures, 74–7
space, polyhedra filling, 50
square centimetre, 93, 94
square metre, 93
squares, 29; comparison of areas by counting, 85–7, 89; in semi-regular tessellations, 49; tessellations from 30, 31, 32; tiling patterns from, 10
standard grids, for measuring area, 94–103
standard units, of length, 60, 62–5
statistics, 234–51
string, area enclosed by loop of, 102, 103
subsets, 21–3
subtraction, of decimals with abacus, 64–9; of fractions, 133
symmetry, lines of, 44, 50; rotational, 12, 13, 42–4, 50, 51; in tiling patterns, 2, 5

tangrams, 112–13
tessellations, 28–52; regular, 32, 33, 48; semi-regular, 46–9; and units of area 84–91
tetrahedra, edges, faces, and vertices of, 25; regular, space-filling with, 50
tetriamonds, 46, 47
tetrominoes, 44, 45
tiling patterns, 1–13
time tables, reference points in, 255–7
topological transformations, 270–81, 303
topology, 270–313
trapezium, 5, 7; tiling patterns from, 11
traversable networks, 288, 291–3, 303, 304
tree diagram, 228
triamonds, 46, 47
triangle numbers, 25
triangles, comparison of areas by counting, 86–91; tessellations from, 30, 31, 34, 40, 41; tiling patterns from, 1–5, 10, 12, 13
trominoes, 44, 45

units, of area, 84–91, 93; of length, 54, 55; of length, fractions of, 55–63; of length, standard, 60, 62–5

vehicle survey, 245–7
vertices of polygons, in semi-regular tessellations, 46, 47, 48, 49
vertices of polyhedra, formula for edges, faces, and, 24, 25, 289

work cards, for tessellations and polyhedra, 52

zero, reference point as, 254, 255, 256, 260, 264